283
Topics in Current Chemistry

Topics in Current Chemistry
Recently Published and Forthcoming Volumes

Anthracycline Chemistry and Biology II

Mode of Action, Clinical Aspects and New Drugs

Volume Editor: Karsten Krohn

With contributions by

F.-M. Arcamone · B. L. Barthel · G. L. Beretta · M. Broggini
D. J. Burkhart · M. L. Capobianco · C. V. Catapano · J.-C. Florent
L. Gianni · B. T. Kalet · T. H. Koch · F. Kratz · P. Menna
G. Minotti · C. Monneret · G. C. Post · D. L. Rudnicki
E. Salvatorelli · F. Zunino

 Springer

The series *Topics in Current Chemistry* presents critical reviews of the present and future trends in modern chemical research. The scope of coverage includes all areas of chemical science including the interfaces with related disciplines such as biology, medicine and materials science. The goal of each thematic volume is to give the nonspecialist reader, whether at the university or in industry, a comprehensive overview of an area where new insights are emerging that are of interest to a larger scientific audience.

As a rule, contributions are specially commissioned. The editors and publishers will, however, always be pleased to receive suggestions and supplementary information. Papers are accepted for *Topics in Current Chemistry* in English.

In references *Topics in Current Chemistry* is abbreviated Top Curr Chem and is cited as a journal.

Visit the TCC content at springerlink.com

ISBN 978-3-540-75812-9   e-ISBN 978-3-540-75813-6
DOI 10.1007/978-3-540-75813-6

Topics in Current Chemistry ISSN 0340-1022

Library of Congress Control Number: 2007936950

Cover design: WMXDesign GmbH, Heidelberg
Typesetting and Production: LE-TEX Jelonek, Schmidt & Vöckler GbR, Leipzig

Printed on acid-free paper

9 8 7 6 5 4 3 2 1 0

springer.com

Topics in Current Chemistry
Also Available Electronically

For all customers who have a standing order to Topics in Current Chemistry, we offer the electronic version via SpringerLink free of charge. Please contact your librarian who can receive a password or free access to the full articles by registering at:

springerlink.com

If you do not have a subscription, you can still view the tables of contents of the volumes and the abstract of each article by going to the SpringerLink Homepage, clicking on "Browse by Online Libraries", then "Chemical Sciences", and finally choose Topics in Current Chemistry.

You will find information about the

– Editorial Board
– Aims and Scope
– Instructions for Authors
– Sample Contribution

at springer.com using the search function.

Color figures are published in full color within the electronic version on SpringerLink.

Preface

More than 15 years have passed since publication of the last monograph on anthracycline antibiotics, the ACS Symposium Series 574, edited by W. Priebe. However, anthracycline antibiotics continue to be one of the most applied antitumor agents, mostly in combination therapy. In addition, a number of exciting new developments such as prodrug development or new synthetic, semi-synthetic, or biosynthetic derivatives have emerged in spite of a certain decrease in synthetic activity. With this background in mind, I accepted the invitation of Prof. J. Thiem to edit an updated collection of reviews on anthracycline antibiotics. In fact, this task turned out to be an exciting endeavor and instead of the initially planned single volume, the numerous contributions from many experts in this exciting field had to be collected into two volumes. The last decade has provided a much greater amount of new information then initially anticipated and these volumes represent a condensed review of this data derived from journals representing quite different fields.

The first volume is dedicated to biological occurrence and biosynthesis as well as the synthesis and chemistry of anthracyclines. Since the pioneering review of H. Brockmann on naturally occurring anthracyclines in 1963, no systematic overview has appeared and this volume will provide a review of the latest information. This topic is closely related to biosynthesis and the intriguing progress in biotechnology to produce biosynthetic anthracycline variants is presented. The part of the volume covering synthesis comprises an updated overview on asymmetric synthesis, combinatorial synthesis using the Diels–Alder reaction, synthesis of fluorinated anthracyclines, the sugar moieties, non-natural glycosyl anthraquinones as DNA binding and photo-cleaving agents, and finally of anthracyclines and fredericamycin A via strong base-induced cycloaddition reaction.

The second volume is devoted to mode of action, clinical aspects, and new drugs. At this point I would like to thank F. M. Arcamone for his invaluable help in selecting the topics and authors of this second volume. Knowledge of the molecular mechanisms of anthracycline activity is of prime importance, also for clinical application, and therefore this is the first contribution of the second volume. The most severe side effect of anthracyclines and many other anticancer drugs is cardiotoxicity, and this has to be given prime importance. Future attempts at reducing this and other side effects include the development of less toxic prodrugs. Therefore, four reviews within this volume are dedicated to this topic: Daunomycin–TFO conjugates for downregulation of gene expression, acid-sensitive prodrugs of doxorubicin, anthracycline–formaldehyde conjugates and their targeted prodrugs, and doxorubicin conjugates for selective delivery to tumors. Last but not least, two chapters are devoted to the recent development of new and hopefully even better anthracycline anticancer drugs: Sabarubicin and nemorubicin. Clinical development of these compounds is approaching and will hopefully give encouraging results.

The two volumes on anthracyclines cover a large area from biotechnology to synthesis and clinical application. Thus, although the chemical aspects dominate, the books will be of value to a broader spectrum of readers looking for recent information on this most important class of antitumor antibiotics.

It has been a great pleasure to work with the competent team of Springer, in particular Dr. Marion Hertel and Birgit Kollmar-Thoni. They have my thanks in addition to all of the authors for their (mostly) timely contributions.

Paderborn, January 2008   Karsten Krohn

Contents

Contents of Volume 282

Anthracycline Chemistry and Biology I

Biological Occurence and Biosynthesis, Synthesis and Chemistry

Volume Editor: Krohn, K.
ISBN: 978-3-540-75814-3

Top Curr Chem (2008) 283: 1–19
DOI 10.1007/128_2007_3
© Springer-Verlag Berlin Heidelberg
Published online: 23 October 2007

Molecular Mechanisms of Anthracycline Activity

Giovanni Luca Beretta · Franco Zunino (✉)

Fondazione IRCCS Istituto Nazionale dei Tumori, Via Venezian 1, 20133 Milan, Italy
franco.zunino@istitutotumori.mi.it

Abstract On the basis of evidence that anthracyclines are DNA intercalating agents and DNA is the primary target, a large number of analogs and related intercalators have been developed. However, doxorubicin and closely related anthracyclines still remain among the most effective antitumor agents. Multiple mechanisms have been proposed to explain their efficacy. They include inhibition of DNA-dependent functions, free radical formation, and membrane interactions. The primary mechanism of action is now ascribed to drug interference with the function of DNA topoisomerase II. The stabilization of the topoisomerase-mediated cleavable complex results in a specific type of DNA damage (i.e., double-strand protein-associated DNA breaks). The drug-stabilized cleavable complex is a potentially reversible molecular event and its persistence, as a consequence of strong DNA binding, may be recognized as an apoptotic stimulus. Indirect evidence supports the notion that the bioreductive processes of the quinone moiety generating the semiquinone radical with concomitant production of reactive oxygen species may contribute to the drug effects. The cellular defense mechanisms and response to genotoxic/cytotoxic stress appear to be critical determinants of the tumor sensitivity to anthracyclines.

Keywords Anthracyclines · Cellular resistance · Cleavable complex · DNA damage · Topoisomerase II

Abbreviations
BCRP   Breast cancer resistant protein
BSO   Buthionine sulfoximine
GSH   Glutathione
LRP   Lung resistant protein

MDA   Malondialdehyde
MRP   Multidrug resistance-associated protein
NAC   *N*-Acetyl cysteine
P-gp   P-glycoprotein
Top II Topoisomerase II
ROS   Reactive oxygen species
SOD   Superoxide dismutase

1
Introduction

Anthracyclines represent a major class of antitumor antibiotics. The most effective member, doxorubicin, is one of the most widely used antitumor agents because of its broad spectrum of antitumor activity. The clinical success of daunorubicin and doxorubicin, the first generation of anthracyclines, has stimulated an intensive effort in the synthesis of analogs or structurally related compounds [1]. In spite of the preclinical development of a large number of agents of this class, only a small number of anthracyclines or related DNA intercalating agents are available for clinical use.

The basic structure of anthracyclines consists of a tetracyclic aglycone linked to an amino sugar (Fig. 1). In an attempt to improve the therapeu-

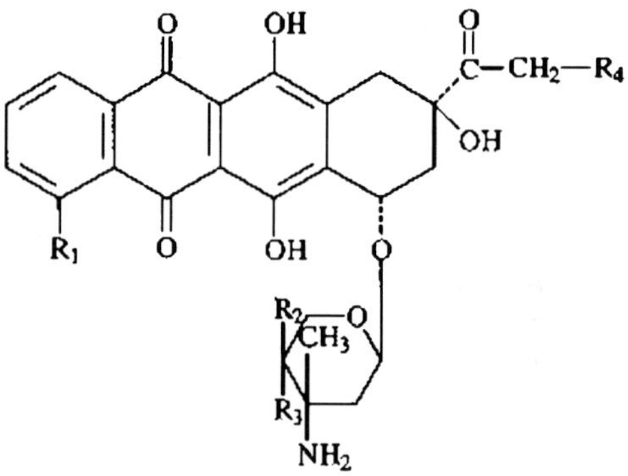

	R_1	R_2	R_3	R_4
Doxorubicin	—OCH$_3$	—H	—OH	—OH
Daunorubicin	—OCH$_3$	—H	—OH	—H
Epirubicin	—OCH$_3$	⸺OH	—H	—OH
Idarubicin	—H	⸺H	—OH	⸺H

Fig. 1 Chemical structure of clinically available anthracyclines

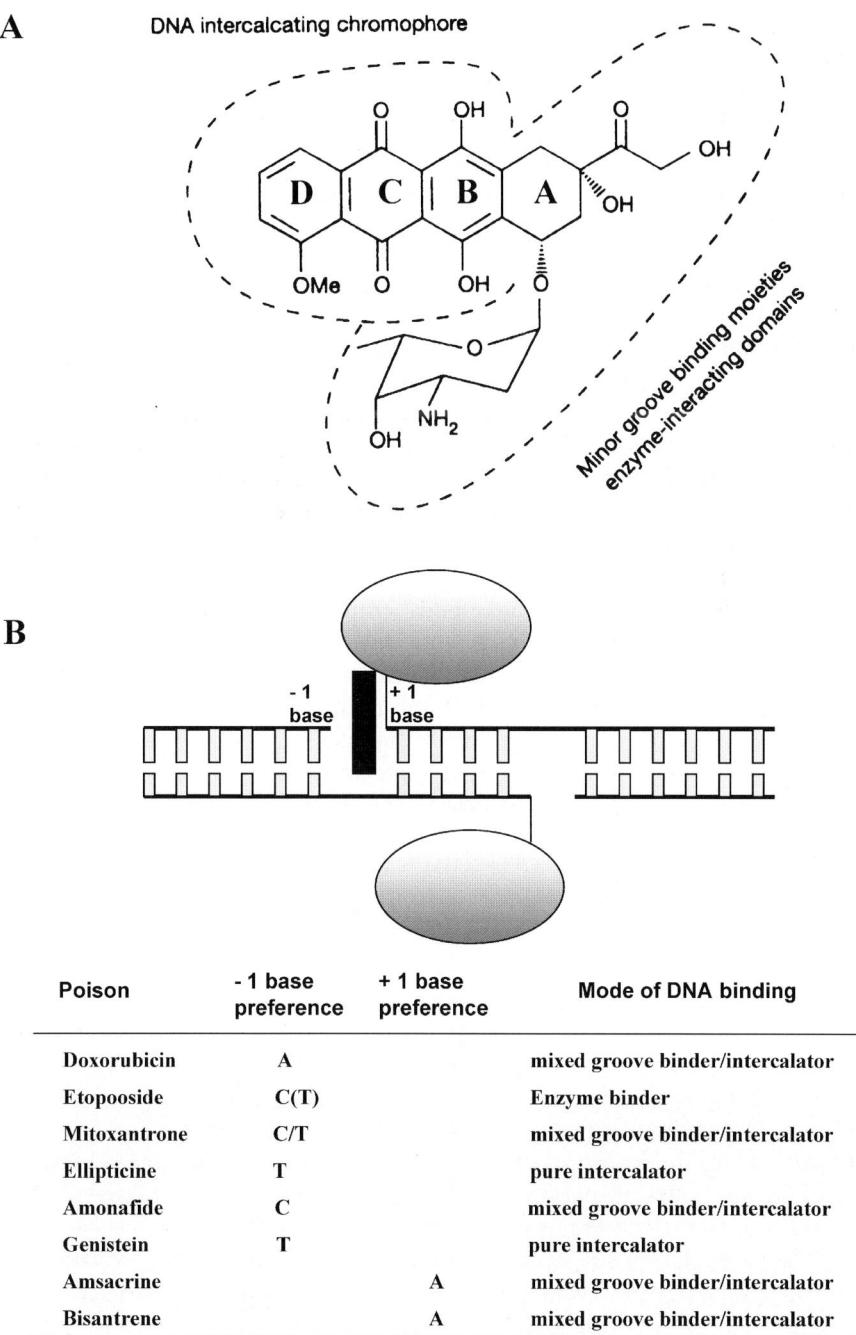

Fig. 2 A Schematic representation of the putative functional moieties of the anthracyclines. **B** Schematic model of the ternary DNA–drug–enzyme complex. The base preferences at the −1/ +1 positions for various effective topoisomerase II poisons are indicated

Poison	−1 base preference	+1 base preference	Mode of DNA binding
Doxorubicin	A		mixed groove binder/intercalator
Etopooside	C(T)		Enzyme binder
Mitoxantrone	C/T		mixed groove binder/intercalator
Ellipticine	T		pure intercalator
Amonafide	C		mixed groove binder/intercalator
Genistein	T		pure intercalator
Amsacrine		A	mixed groove binder/intercalator
Bisantrene		A	mixed groove binder/intercalator

tic and pharmacological properties of the natural compounds, a number of modifications of the basic structure were taken into consideration, including changes or substitutions of the C-9 side chain, the amino sugar, or the aglycone moiety [1]. Indeed, each of these moieties has been implicated in critical interactions with the cellular target (Fig. 2). No analog to date has shown an activity clearly superior to that of doxorubicin, which remains the best anthracycline [2].

Daunorubicin is one of the most effective agents in the treatment of acute lymphocytic and myelogenous leukemias, but it has little activity against solid tumors. Doxorubicin exhibited a broad spectrum of activity and remains one of the most effective drugs for the treatment of solid tumors. Breast carcinoma, small-cell lung carcinoma, and ovarian carcinoma are the most doxorubicin-responsive solid tumors. Epirubicin, which is characterized by epimerization of the hydroxyl group at position 4 of the amino sugar (Fig. 1), exhibits an improved toxicological profile and reduced cardiotoxicity with antitumor efficacy and spectrum of activity similar to that of doxorubicin. Idarubicin, which lacks a methoxy group in position 4 of the chromophore, shows an enhanced lipophilicity and changed pharmacological profile. In comparison to daunorubicin, the removal of the methoxy group markedly enhances the drug's ability to induce topoisomerase II (Top II)-mediated DNA cleavage [3] as well as an increased antitumor potency as a consequence of an enhanced intracellular drug accumulation. The drug lipophilicity allows oral absorption, providing an additional advantage of idarubicin over daunorubicin which is completely inactive when administrated orally. The clinical efficacy of idarubicin is restricted to the treatment of leukemia, since a lower activity against solid tumors is a common feature of daunorubicin analogs.

2
Anthracyclines as Intercalating Agents

Doxorubicin, like other effective anthracycline glycosides, is a well-known DNA intercalating agent and DNA is recognized as being the primary target for its pharmacological action. The drug–DNA intercalation complex, resulting from the insertion of the planar tetracyclic chromophore between adjacent base pairs, is stabilized by electrostatic interactions between DNA phosphate groups and the positively charged amino group of the sugar moiety. The intercalation site depends not only on the planar chromophore, but also on a variety of intrinsic properties (steric and electronic) that also involve the external binding moieties (Fig. 2). However, the cytotoxic activity is not simply related to the drug's ability to bind to DNA, since the mode and the site of binding appear to be more critical than the binding affinity. Although DNA binding is central to the antitumor activity of anthracyclines, available

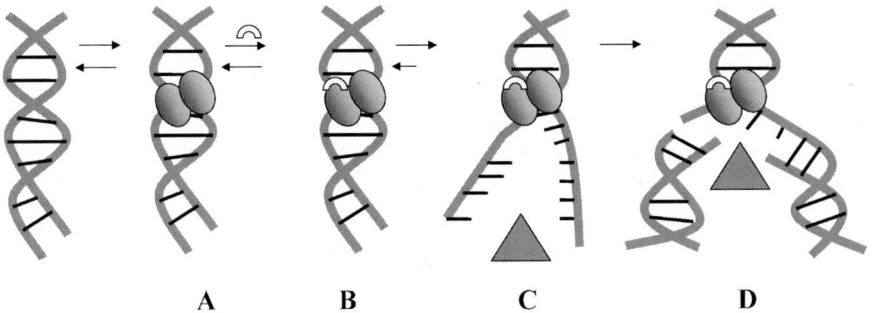

Fig. 3 Mechanism of DNA topoisomerase II poisoning. **A** Cleavage reaction of topoisomerase II. **B** Drug-stabilized cleavable complex. **C** Collision of the replication fork. **D** Irreversible double-strand break

evidence indicates that it is the inhibition of a specific DNA function that is responsible for their therapeutic effects.

Indeed, the primary mechanism of cytotoxic and antitumor activity of effective intercalating agents is now ascribed to their interference with the function of Top II [3, 4]. Antitumor inhibitors of topoisomerases function as enzyme poisons by forming a DNA–drug–enzyme ternary complex, thus stabilizing the cleavable complex, in which DNA strands are broken and enzyme subunits are covalently linked to DNA [5]. It is likely that the intercalating agent is placed at the interface between the enzyme active site and the DNA cleavage site, thus preventing DNA religation. Stabilization of the cleavable complex causes specific lethal DNA damage (i.e., double-strand protein associated DNA breaks) after collision with enzymes involved in DNA metabolism (Fig. 3). Additional details of the topoisomerase-mediated mechanism are described in the following section.

3
Anthracyclines as DNA Damaging Agents

Multiple mechanisms of DNA damage produced by anthracyclines have been described. Although the Top II poisoning is the best characterized, other mechanisms have been proposed including free radical formation. In particular, the semiquinone radical can intercalate with DNA resulting in DNA damage, and reactive oxygen species (ROS) are known to induce DNA damage. The interaction with the DNA–Top II complex is likely a primary event for growth arrest and/or cell killing at pharmacologically relevant concentrations. Free radical formation could contribute to cytotoxic effects [6], but their effect can only be detected at high drug levels.

3.1
DNA Damage Mediated by Topoisomerase II Poisons

DNA topoisomerases are nuclear enzymes that regulate DNA topology during multiple DNA functions (including transcription, replication, and recombination), and are essential for the integrity of genetic material [5]. During its catalytic function, the cleavage of the DNA leads to the formation of a Top II–DNA covalent complex (cleavable complex) that occurs through the tyrosine hydroxyl groups of the enzyme and the 5′ terminus of the DNA. By restoring the original unbroken DNA form, Top II generates reaction products whose chemical structures are indistinguishable from those of its substrates, because only the spatial orientation of the genetic material is changed by the reaction [7, 8].

Top II is a homodimeric enzyme which requires ATP as a cofactor for the catalytic activity [9]. Mammalian cells have two type II isozymes, termed Top IIα (170 kDa) and Top IIβ (180 kDa). The two isozymes differ in their patterns of expression, with Top IIα preferentially expressed in proliferating cells while Top IIβ is apparently expressed at equal levels in proliferating and quiescent cells. The two enzymes show considerable amino acid conservation throughout most of the protein coding region and share identical enzymatic properties [10]. Top II is able to remove negative or positive superhelical twists from DNA and resolve intramolecular DNA knots as well as intermolecular tangles [11]. The enzyme passes one duplex strand of DNA, designated as the transport, or "T-segment", through a transient double-strand break into another duplex DNA, designated as the DNA gate, or "G-segment". In the two-gate model, the T-segment enters the enzyme through a gate on one side of the protein, is transported through the G-segment located within the enzyme, and then exits through a second protein gate on the other side of the enzyme [12]. The four steps of the catalytic cycle of Top II are shown in Fig. 4.

Step 1, DNA binding. Top II initiates its catalytic cycle by binding to the DNA substrate, the G-segment. This interaction requires no cofactor, although stimulation of binding has been reported in the presence of divalent cations [13] and is governed by two features of the double helix: nucleotide sequence and topological structure. Top II binds to DNA at preferred nucleic acid sequences which define the site of enzyme-mediated DNA cleavage [14].

Step 2, pre-strand passage. The enzyme generates double-strand breaks by making two coordinate nicks [15] on opposite strands of the G-segment and leaves a four-base 5′-overhang on each cleaved nucleic acid strand [16, 17] (Fig. 2). During scission, Top II forms covalent bonds between its tyrosyl residue (one per monomer) and the newly generated 5′-phosphate termini of the DNA [15]. This DNA cleavage intermediate is referred to as the "cleavable complex". Top II cleaves DNA at preferred sequences, but the stringency of sequence recognition is low and probably reflects the fact that the enzyme

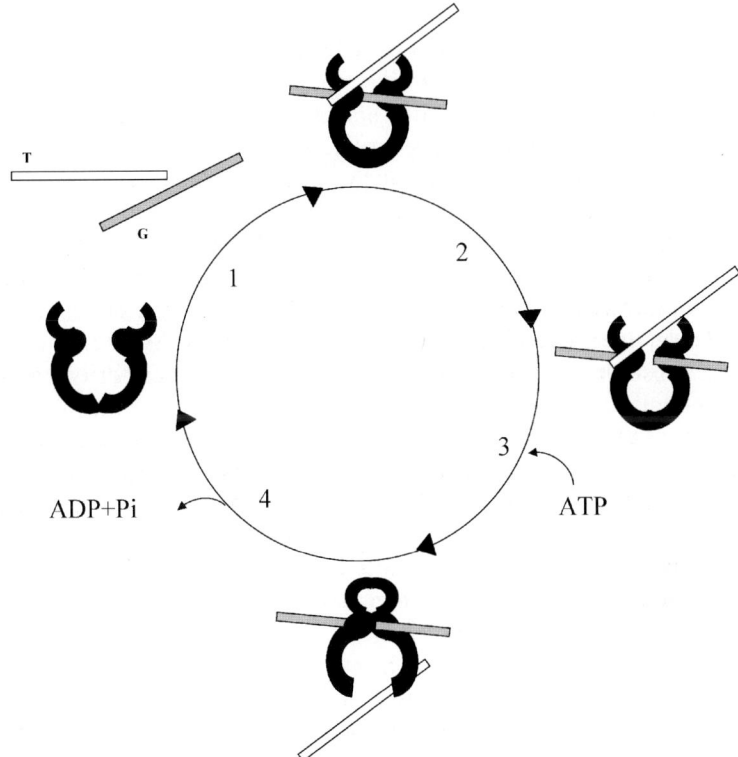

Fig. 4 Catalytic cycle of DNA topoisomerase II

functions in a global manner and therefore is not restricted to a small number of unique sites in the genome.

Step 3, DNA strand passage. Upon binding of its ATP cofactor, Top II undergoes a conformational change that traps the second DNA segment, the T-segment, and triggers double-stranded DNA passage [9]. The T-segment gets passed through a double-strand break made within the G-segment.

Step 4, ATP hydrolysis. Top II hydrolyzes the high energy cofactor, which triggers the opening of the protein clamp and releases nucleic acid products [13, 18]. The enzyme regains the ability to initiate a new round of catalysis.

In the presence of a poison, which is able to affect the DNA cleavage–religation process, the cleavage complex has a longer life and the persistence of DNA breaks depends on the drug interaction in the ternary complex [16, 19]. This action converts Top II into an endogenous toxin that produces DNA damage and triggers a series of cellular events, resulting in cell cycle arrest and/or cell death.

Some poisons do not bind to DNA, while others exhibit a high affinity for the nucleic acid. These include pure intercalators and mixed groove binder/intercalators which are the most effective inhibitors, characterized by a planar polycyclic DNA-intercalating system to which side chain groups are eventually attached. The planar moiety intercalates between DNA bases producing efficient stacking interactions and the side chain interacts with DNA minor groove and with the enzyme, thus acting as enzyme-recognition elements (Fig. 2). The relative position of the (planar ring/side chain) pharmacophores plays a major role in modulating nucleic acid binding and enzyme poisoning effects [5, 20].

In the ternary complex, the drug is bound to the DNA and the enzyme simultaneously [21]. Hence, there are two pharmacophoric regions in a poison molecule, one interacting with the enzyme and the other with the nucleic acid. The poisoning of the cleavable complex by anticancer drugs does not occur randomly along the DNA chain [22]. Top II activity on the genome exhibits per se a certain degree of specificity and significant nucleotide preferences are observed in the regions flanking the cleavage site, often corresponding to alternating purine–pyrimidine tracts [22, 23]. However, site selectivity exhibited by the poisons is more efficient and dependent on the nature of the poison. Compounds of different chemical classes stimulate specific cleavage patterns in DNA fragments, which do not comprise all of the sites recognized by the enzyme alone. Moreover, the effects of drug specificity are principally due to the base immediately preceding (–1) or following (+1) the cleavage site [20] (Fig. 2). Generally, drugs having a strong –1 preference do not exhibit a strong +1 preference, indicating that effective recognition of DNA by a drug is carried out either on one side or on the other of the cleavage complex, and that the drug molecule can localize either upstream or downstream of the cleaved phosphodiester bond. A purine residue is invariably found to be preferred at +1, while the majority of –1 specific agents accept both pyrimidines and purines in the cut strand.

The available Top II poisons exhibit a low level of discrimination between the two isoenzymes. Like acridines, anthraquinones, epipodophyllotoxins, and ellipticine [24, 25], anthracyclines are able to affect both the α and the β isoforms, although to different extents, indicating that both enzymes are potential targets. It appears that the anthracyclines prefer the α enzyme, and amsacrine and mitoxantrone the β enzyme. Relevant to this point is the recent finding that the two isoenzymes of Top II have a different role in antitumor therapy [26]. Indeed, Top IIα appears to be implicated in tumor cell killing and Top IIβ in the development of treatment-related secondary tumors. Therefore, the preference of anthracyclines for the α isoform supports potential therapeutic advantages over other Top II inhibitors.

As a consequence of their high DNA binding affinity, at sufficiently high concentrations, anthracyclines are also able to inhibit the catalytic activity of the enzyme without stimulating DNA cleavage, whereas at low concentrations they stimulate cleavage [3, 20, 27].

3.2
DNA Damage Mediated by Reactive Oxygen Species

The potential involvement of free radical generation in the cytotoxicity of the anthracyclines is complex and not completely understood. There is no question that under the appropriate conditions the chemical nature of the anthracyclines leads to the generation of reactive free radicals. The unresolved question is whether free radicals are generated at pharmacologically relevant concentrations of the anthracyclines, and whether such free radicals could contribute to their antitumor efficacy and/or toxicity [6].

The quinone moiety of anthracyclines can undergo reduction to a hydroquinone form, with formation of intermediate semiquinone free radicals which can reduce oxygen to produce superoxide and other ROS, including hydrogen peroxide and hydroxyl radicals [28] (Fig. 5). The reactive species could induce DNA damage and lipid peroxidation. The precise role of free radical formation is still a matter of debate. The DNA damage, which, unlike that associated with to Top II poisons, is not protein-associated [29], is blocked by superoxide

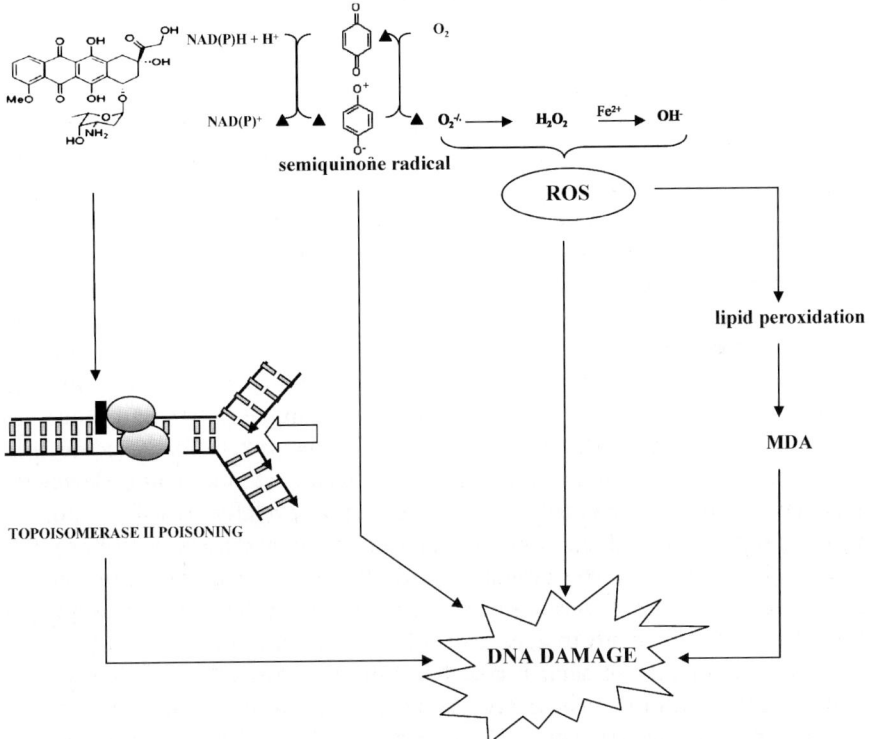

Fig. 5 Anthracyclines and DNA damage. A schematic representation of DNA damage mediated by ROS is shown

dismutase (SOD) and catalase. It is generally believed that appreciable levels of free radicals, generated by redox metabolism of anthracyclines, are produced at elevated drug concentrations, in contrast to protein-associated strand breaks alone (resulting from inhibition of Top II) which are evident at lower (pharmacologically relevant) drug concentrations [30].

ROS produced by anthracyclines induce lipid peroxidation with formation of malondialdehyde (MDA) (Fig. 5). Like other enals, MDA reacts at the exocyclic amino groups of deoxyguanosine, deoxyadenosine, and deoxycytidine to form alkylated products [31]. MDA is mutagenic in human cells, with the majority of MDA-induced mutations occurring at GC base pairs and consisting of large insertions and deletions [32]. In proliferating cells the formation of MDA–DNA adducts is accompanied by cell cycle arrest and inhibition of cyclin E- and cyclin B-associated kinase activities in both wild-type p53 and p53-null cell lines [33].

Several lines of evidence support the idea that superoxide anion and hydrogen peroxide may be implicated in influencing cell proliferation or cell death, depending on the extent of oxidative stress [34]. In addition, the apoptosis pathway activated by DNA damage may involve the formation of ROS and oxidative damage of mitochondrial components [35]. Thus, the possibility that anthracycline-mediated oxidative stress may be a contributing factor to drug-induced cell death cannot be ruled out.

A contribution of ROS in the mechanism of cytotoxicity is supported by the effects of scavenger agents. Indeed, many scavenger molecules (e.g., glutathione (GSH), N-acetyl cysteine (NAC), GSH reductase, SOD, catalase) are known to reduce the cytotoxic effects of anthracyclines [36]. Among these, GSH is the most abundant non-protein thiol in cells (being found in the millimolar range in most tissues), and is a major component of the process for defense against the toxicity of xenobiotics and oxidants. The major pathway for GSH metabolism in defense of the cell is the reduction of hydrogen peroxide and lipid hydroperoxides by GSH peroxidases. The depletion of GSH and GSH reductase by 1,3-bis(2-chloroethyl)-1-nitrosourea (BCNU) resulted in increased lipid peroxidation by doxorubicin [37]. In addition, resistance to doxorubicin in a subline of HL-60 is reversed by the GSH depleting agent buthionine sulfoximine (BSO) [38]. The interpretation that ROS provide a contribution to the cytotoxic activity of anthracyclines is also supported by the observation that, among intercalating Top II inhibitors, anthracyclines and anthraquinones (containing a quinone moiety in their structure) are the most effective as antitumor agents. Again, scavengers of ROS have been reported to protect cells from apoptosis induced by other Top II inhibitors.

Whereas available evidence supports the view that the primary mechanism of antitumor action is related to Top II inhibition, the drug redox metabolism has been implicated in cardiotoxicity [39, 40]. It is thus conceivable that antitumor activity and cardiotoxicity occur through different biochemical mechanisms. The possibility is supported by some indirect evidence: (1) free radical scavengers or antioxidant agents (e.g., tocopherol, NAC) can reduce cardiac toxicity without affecting tumor inhibition in pre-clinical systems; (2) 5-imino-daunorubicin, which is markedly less effective in producing reductase-dependent free radicals, lacks cardiac toxicity but retains some antitumor activity; and (3) the biochemical profile of cardiac tissue in terms of ability to protect against oxidative damage (high levels of reductase and low activity of catalase and SOD) could account for the organ-specific toxicity of anthracyclines.

4
Mechanisms of Cell Death

The stabilization of the cleavable complexes is the primary cytotoxic lesion, but the outcome of treated cells depends on a number of downstream events, including (a) the processing of stabilized cleavage complexes into lethal DNA double-strand breaks; (b) response to the DNA damage, leading to activation of stress-associated signaling pathways and cell cycle arrest; and (c) activation of cell death pathways.

(a) Processing of Cleavage Complexes into Cytotoxic DNA Double-Strand Breaks

The conversion of Top II–DNA complexes into cytotoxic damage is well un-derstood at the molecular level. Reversible Top II–DNA–drug cleavage com-plexes are converted into irreversible DNA double-strand breaks through col-lision with enzymes involved in DNA replication and transcription (e.g., DNA helicases) [41] (Fig. 3). Indeed, agents that inhibit DNA helicases diminish the cytotoxicity of drugs [42, 43] and inhibitors of replication or transcription at-tenuate the cytotoxic effects of Top II poisons [44–46]. In addition, Top II poisons appear to be particularly toxic during S phase, when replication forks and transcription complexes are both present, as compared to other phases of the cell cycle [45, 47].

(b) Response to the DNA Damage

The cytotoxic lesions generated from the cleavage complex induce DNA-damage signaling pathways, including cell cycle arrest and activation of DNA repair processes (Fig. 6).

Anthracyclines, like many other genotoxic agents, activates a p53-mediated response. On the basis of the crucial role of p53 in activation of apoptosis, p53 could play an important function in anthracycline cytotoxicity. Preclin-ical and clinical studies support this interpretation [48–50] but conflicting results have been reported [51, 52]. Uncertainties about the role of p53 in anthracycline-induced apoptosis may be attributed to such various factors

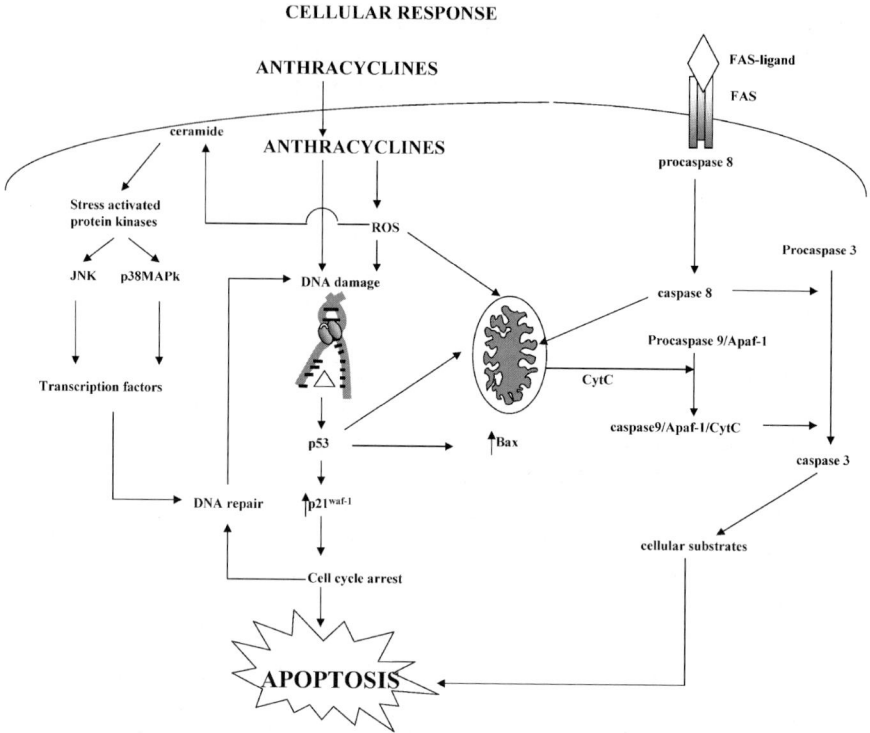

Fig. 6 Cellular response to anthracyclines

as heterogeneity of the tumors or of the methods used for assessing p53 status and tumor response [53]. Anthracycline-dependent p53 activation contributes to the activation of p21^{wafl}, an inhibitor of cyclin-dependent kinases involved in DNA damage checkpoints. Cyclin-dependent kinases catalyze the highly orchestrated events that drive the cell through the cell cycle. Inhibition of these kinases has been associated with arrest of drug-treated cells in G_1 and/or G_2 phases of the cell cycle. Although this mechanism has been proposed to contribute to G_1 arrest of p53-proficient cells, it has also been suggested that p21^{wafl} expression might protect cells from anthracyclines because the G_1 block facilitates DNA repair before the cells undergo DNA replication. Indeed, constitutively high levels of p21^{wafl} have been associated with chemoresistance in acute myelogenous leukemia [54].

(c) Activation of Cell Death Signals

Common responses to DNA damage are cell cycle arrest and apoptosis. Usually, DNA damaging agents, including Top II inhibitors, cause cell cycle arrest in G_2. The signals activated by topoisomerase-mediated DNA cleavage (likely) involve p53, which have multiple functions, including activation of cell cycle checkpoints and DNA repair processes and stimulation of apoptosis. Thus, the cell outcome is likely determined by a balance between proapoptotic and protective signals. Anthracyclines are known to activate both the intrinsic and extrinsic pathways of apoptosis; p53 function is likely implicated in regulation of both pathways.

Friesen and coworkers reported that treatment of CEM human leukemia cells with a variety of agents, including anthracyclines, results in enhanced expression of Fas ligand [55]. Interestingly, drug-induced apoptosis was diminished in CEM cells selected for resistance to Fas-mediated cell death [55], suggesting that alterations in the Fas/Fas ligand signaling pathway might constitute a mechanism of resistance to these agents. Similar results were observed in neuroblastoma cell lines treated with doxorubicin or etoposide [56]. The Fas/Fas ligand apoptotic pathway is initiated by the proximity-induced activation of caspase-8 and subsequent activation of downstream proteases [57, 58], resulting in cell death (Fig. 6). This pathway could be activated by a drug-induced increase of the lipid second messenger ceramide [59]. Ceramide has been shown to activate a limited number of protein kinases, including stress-activated protein kinases [59, 60]. The stress-activated kinase pathway consists of a cascade of cytoplasmic kinases, including JNK and p38 [61, 62], which ultimately leads to phosphorylation of transcription factors (e.g., c-jun) that activate transcription of stress-responsive genes, which undergo apoptotic cell death (Fig. 6) [62].

Moreover, it is evident that anthracyclines can directly induce the release of cytochrome c from mitochondria, thereby inducing apoptosis regardless of DNA damage or signaling pathways or p53 status [63, 64].

5
Mechanisms of Cellular Resistance

As with other cytotoxic drugs, clinical resistance to anthracyclines may be a multifactorial phenomenon likely involving pharmacological and tumor-related factors. Based on preclinical studies, cellular alterations that could contribute to resistance to anthracyclines include (1) pretarget events (i.e., drug accumulation, detoxification, and intracellular drug distribution), which result in inadequate accumulation/subcellular localization of drug in the cells; (2) target-related events including quantitative/qualitative alterations of the target, e.g., reduced drug–target interaction as a consequence of enzyme downregulation or Top II gene mutation; and (3) posttarget events, i.e., alterations in the cellular response to DNA damage generated by the formation of the ternary complex (Fig. 7).

Anthracyclines, owing to modest lipophilicity and net positive charge at physiological pH, are efficient substrates for energy-dependent drug efflux pumps, such as P-glycoprotein (P-gp), multidrug resistance-associated pro-

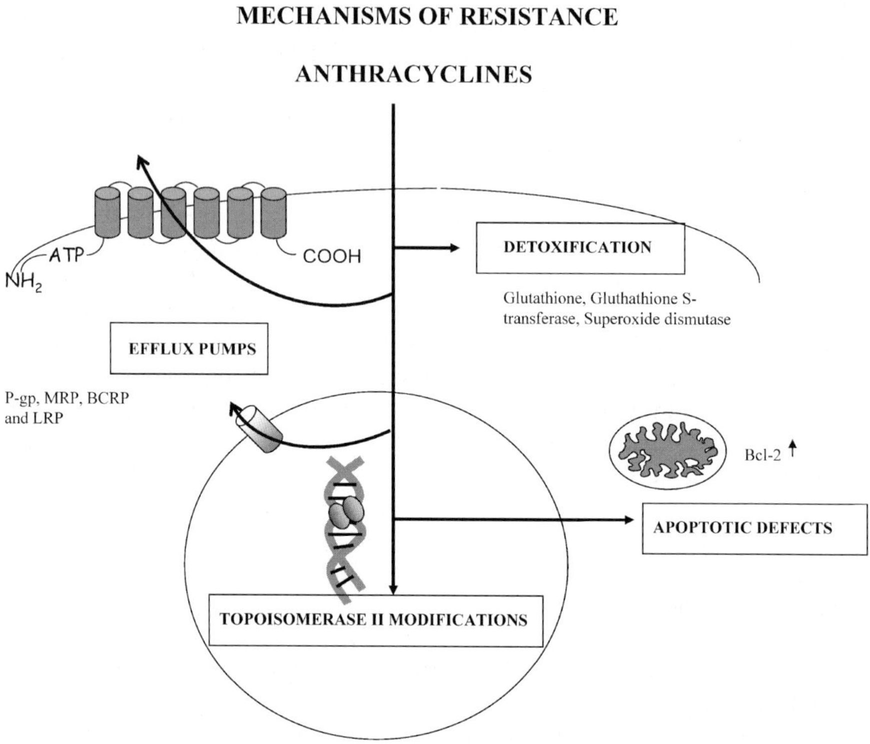

MECHANISMS OF RESISTANCE

ANTHRACYCLINES

Fig. 7 Mechanisms of cellular resistance to anthracyclines

tein (MRP) [65], breast cancer resistant protein (BCRP) [66], and lung re-sistant protein (LRP) [67]. P-gp is not tumor-specific, since it is expressed also in normal tissues, including kidney, liver, colon, and adrenal gland, and is involved in the transport of hydrophobic metabolites or hormones. Thus, expression of P-gp in tumors derived from such tissues could account for their natural resistance. A multidrug-resistant phenotype has been found to be associated with the overexpression of MRP [68]. Unlike P-gp, which ap-pears to transport unmodified drugs and xenobiotics, BCRP, a member of the subfamily G of the human ABC superfamily, is an organic anion pump able to transport conjugates of sulfates, GSH, and glucuronic acid and is very efficient in transporting GSH conjugates [69]. Overexpression of transport systems may also play a role in subcellular sequestration of drugs (thus reduc-ing the drug–target interaction). A similar function has been proposed for the human major vault protein LRP [67]. Mutations in amino acid 482 of BCRP may occur in cells selected for resistance to doxorubicin, and this residue has been implicated in substrate interaction [66].

Resistant cells selected following continuous exposure to doxorubicin may express high levels of GSH, metallothionein, glutathione S-transferase, Cu, Zn superoxide dismutase, or Mn superoxide dismutase [70–72]. These factors are implicated in detoxification processes and in cell protection against oxidative stress.

Quantitative (level of expression) or qualitative (e.g., mutation) alterations of the target enzyme are expected to influence the drug's ability to induce an adequate extent of DNA lesions [73]. Reduced Top II expression or specific alterations in enzyme activity or sensitivity may be responsible for a different multidrug resistance phenotype which involves only Top II poisons.

Despite adequate drug concentration at the target level, the cellular response to DNA damage may play a critical role in determining chemosensitivity and may account for the heterogeneity of tumor response to drug treatment. Defects in cell cycle checkpoints and apoptotic signaling pathways may critically influence anthracycline cytotoxicity. Additional studies are needed to determine the clinical relevance of the resistance mechanisms detected in tumor cell systems.

6
Concluding Remarks

Anthracyclines remain among the most clinically effective antitumor agents. The efficacy of the two prototype anthracyclines has stimulated an intensive effort in elucidation of their mechanism of action in an attempt to provide a rational basis for the design of more effective analogs. In spite of this effort, no analog has shown an activity clearly superior to that of doxorubicin. Similarly, among a large variety of intercalating agents, only a few compounds exhibit appreciable activity. The molecular basis of their efficacy and selectivity toward specific tumor types is not understood. The persistence of DNA lesions, as a consequence of the strong intercalation, may be recognized as an apoptotic stimulus. Anthracyclines exhibit a unique sequence specificity of stimulation of enzyme-mediated DNA cleavage. It is conceivable that the drug's ability to damage critical genomic sites may play a role in antitumor activity and in variable responsiveness observed in different tumor types, which is not simply related to the Top II inhibition. The initial DNA lesion, the drug-stabilized cleavable complex, is a potentially reversible molecular event. It is likely that the persistence of DNA damage and not solely its extent is a critical determinant of drug efficacy. For example, doxorubicin is more effective than daunorubicin against solid tumors. Indeed, an increased retention of doxorubicin by tumor cells is expected to cause more persistent DNA damage resulting in an effective tumor response. A plausible explanation of the different behavior of doxorubicin and daunorubicin could be the marked protein binding of doxorubicin. The side chain of doxorubicin could undergo rearrangements between the 13-carbonyl atom and the adjacent carbon atom, resulting in a reversible conversion of the hydroxyketone to the hydroxyaldehyde. The carbonyl group at C-14 of the aldehyde form is capable of reacting with free amino groups of proteins through reversible Schiff's base linkages [74].

Anthracyclines have potential for multiple cellular effects that may be relevant for their cytotoxic and antitumor action. They include DNA interaction, involvement in redox metabolism, and direct membrane effects. The relative relevance of the multiple effects is still a matter of debate. A major problem concerning the interpretation of the relative contribution of various cellular effects is the difficulty in establishing a quantitative correlation between the extent of cellular injury and various parameters of pharmacological activity. Central to the biological effects of anthracyclines is the DNA damage mediated by inhibition of Top II function. A large number of agents, in particular several DNA intercalating agents (e.g., amsacrine, ellipticines), have been described as Top II inhibitors. However, only quinone-containing inhibitors have a relevant antitumor activity. This observation suggests that generation of ROS, which occurs in quinine-containing inhibitors, provides a contribution to the drug cell-killing activity, and indirect evidence supports this interpretation. The intracellular level of GSH, a major component of cellular antioxidant defense, is known to confer resistance to doxorubicin [75]. Several lines of evidence support the notion that ROS may be involved in the regulation of diverse cellular functions, including apoptosis induced by DNA damaging agents [76]. Therefore, in addition to the production of ROS generated by redox metabolism of the quinone moiety, oxidative stress, a consequence of the apoptotic stimulus at the mitochondrial level, may provide a contribution to the biological activity.

Thioredoxin has been reported to enhance the apoptotic death of breast carcinoma cells in response to daunorubicin as a consequence of redox cycling of the drug [77]. The redox cycling process should facilitate the semiquinone intercalation with DNA. If the bioreductive processes generating free radicals (superoxide anion and semiquinone radical) play a relevant role in cytotoxicity of anthracyclines, it is conceivable that the biological background of various tumor types (in particular, the expression of enzymes implicated in redox metabolism) is a critical determinant of cell sensitivity and tumor response to drug treatment.

Whatever is the mechanism of DNA damage, it is now evident that drug efficacy is markedly dependent on the persistence of drug effects. The different cellular pharmacokinetics likely accounts for the increased efficacy of doxorubicin over daunorubicin.

References

1. Arcamone F (1981) Doxorubicin: anticancer antibiotics. Medicinal Chemistry Series, vol 17. Academic, New York
2. Weiss RB (1992) Semin Oncol 19:670
3. Capranico G, Zunino F (1992) Eur J Cancer 28:2055

4. Zunino F, Capranico G (1997) Sequence-selective groove binders. In: Tericher B (ed) Cancer therapeutics: experimental and clinical agents. Humana, Totowa, p 195
5. Capranico G, Binaschi M, Borgnetto ME, Zunino F, Palumbo M (1997) Trends Pharmacol Sci 18:323
6. Gewirtz DA (1999) Biochem Pharmacol 57:727
7. Liu LF, Liu CC, Alberts BM (1980) Cell 19:697
8. Watt PM, Hickson ID (1994) Biochem J 303:681
9. Lindsley JE, Wang JC (1991) Proc Natl Acad Sci USA 88:10485
10. Turley H, Comley M, Houlbrook S, Nozaki N, Kikuchi A, Hickson ID, Gatter K, Harris AL (1997) Br J Cancer 75:1340
11. Wang CJ (1996) Annu Rev Biochem 65:635
12. Roca J, Wang JC (1994) Cell 77:609
13. Osheroff N (1986) J Biol Chem 261:9944
14. Thomsen B, Bendixen C, Lund K, Andersen AH, Sorensen BS, Westergaard O (1990) J Mol Biol 215:237
15. Zechiedrich EL, Christiansen K, Andersen AH, Westergaard O, Osheroff N (1989) Biochemistry 28:6229
16. Liu LF, Rowe TC, Yang L, Tewey KM, Chen GL (1983) J Biol Chem 258:15365
17. Muller MT, Spitzner JR, DiDonato JA, Mehta VB, Tsutsui K, Tsutsui K (1988) Biochemistry 27:8369
18. Miller KG, Liu LF, Englund PT (1981) J Biol Chem 256:9334
19. Sander M, Hsieh T (1983) J Biol Chem 258:8421
20. Capranico G, Binaschi M (1998) Biochim Biophys Acta 1400:185
21. Pommier Y, Pourquier P, Fan Y, Strumberg D (1998) Biochim Biophys Acta 1400:83
22. Lee MP, Sander M, Hsieh T (1989) J Biol Chem 264:21779
23. Svejstrup JQ, Christiansen K, Andersen AH, Lund K, Westergaard O (1990) J Biol Chem 265:12529
24. Guano F, Pourquier P, Tinelli S, Binaschi M, Bigioni M, Animati F, Manzini S, Zunino F, Kohlhagen G, Pommier Y, Capranico G (1999) Mol Pharmacol 56:77
25. Errington F, Willmore E, Tilby MJ, Li L, Li G, Li W, Baguley BC, Austin CA (1999) Mol Pharmacol 56:1309
26. Azarova AM, Lyu YL, Lin C-P, Tsai Y-C, Lau JY-N, Wang JC, Liu LF (2007) Proc Natl Acad Sci USA 104:11014
27. Pommier Y (1993) Cancer Chemother Pharmacol 32:103
28. Myers C (1998) Semin Oncol 25:10
29. Potmesil M, Israel M, Silber R (1984) Biochem Pharmacol 33:3137
30. Fornari FA, Randolph JK, Yalowich JC, Ritke MK, Gewirtz DA (1994) Mol Pharmacol 45:649
31. Marnett LJ, Riggins JN, West JD (2003) J Clin Invest 111:583
32. Niedernhofer LJ, Daniels JS, Rouzer CA, Greene RE, Marnett LJ (2003) J Biol Chem 278:31426
33. Ji C, Rouzer CA, Marnett LJ, Pietenpol JA (1998) Carcinogenesis 19:1275
34. Burdon RH (1995) Free Radic Biol Med 18:775
35. Polyak K, Xia Y, Zweier JL, Kinzler KW, Vogelstein B (1997) Nature 389:300
36. Schisselbauer JC, Crescimanno M, D'Alessandro N, Clapper M, Toulmond S, Tapiero H, Tew KD (1989) Cancer Commun 1:133
37. Babson JR, Abell NS, Reed DJ (1981) Biochem Pharmacol 30:2299
38. Raghu G, Pierre-Jerome M, Dordal MS, Simonian P, Bauer KD, Winter JN (1993) Int J Cancer 53:804
39. Weijl NI, Cleton FJ, Osanto S (1997) Cancer Treat Rev 23:209

40. Minotti G, Menna P, Salvatorelli E, Cairo G, Gianni LG (2004) Pharmacol Rev 56:185
41. Bachur NR, Yu F, Johnson R, Hickey R, Wu Y, Malkas L (1992) Mol Pharmacol 41:993
42. Jensen PB, Sorensen BS, Demant EJ, Sehested M, Jensen PS, Vindelov L, Hansen HH (1990) Cancer Res 50:3311
43. Adjei AA, Charron M, Rowinsky EK, Svingen PA, Miller J, Reid JM, Sebolt-Leopold J, Ames MM, Kaufmann SH (1998) Clin Cancer Res 4:683
44. Holm C, Covey JM, Kerrigan D, Pommier Y (1989) Cancer Res 49:6365
45. D'Arpa P, Beardmore C, Liu LF (1990) Cancer Res 50:6919
46. Kaufmann SH (1991) Cancer Res 51:1129
47. Estey E, Adlakha RC, Hittelman WN, Zwelling LA (1987) Biochemistry 26:4338
48. Penault-Llorca F, Cayre A, Bouchet Mishellany F, Amat S, Feillel V, Le Bouedec G, Ferriere JP, De Latour M, Chollet P (2003) Int J Oncol 22:1319
49. Ruiz-Ruiz C, Robledo G, Cano E, Redondo JM, Lopez-Rivas A (2003) J Biol Chem 278:31667
50. Stearns V, Singh B, Tsangaris T, Crawford JG, Novielli A, Ellis MJ, Isaacs C, Pennanen M, Tibery C, Farhad A et al (2003) Clin Cancer Res 9:124
51. Perego P, Corna E, De Cesare M, Gatti L, Polizzi D, Pratesi G, Supino R, Zunino F (2001) Curr Med Chem 8:31
52. Gariboldi MB, Ravizza R, Riganti L, Meschini S, Calcabrini A, Marra M, Arancia G, Dolfini E, Monti E (2003) Int J Oncol 22:1057
53. Bertheau P, Plassa F, Espie M, Turpin E, de Roquancourt A, Marty M, Lerebours F, Beuzard Y, Janin A, de The H (2002) Lancet 360:852
54. Zhang W, Kornblau SM, Kobayashi T, Gambel A, Claxton D, Deisseroth AB (1995) Clin Cancer Res 1:1051
55. Friesen C, Herr I, Krammer PH, Debatin KM (1996) Nat Med 2:574
56. Fulda S, Sieverts H, Friesen C, Herr I, Debatin KM (1997) Cancer Res 57:3823
57. Medema JP, Scaffidi C, Kischkel FC, Shevchenko A, Mann M, Krammer PH, Peter ME (1997) EMBO J 16:2794
58. Muzio M, Stockwell BR, Stennicke HR, Salvesen GS, Dixit VM (1998) J Biol Chem 273:2926
59. Laurent G, Jaffrezou JP (2001) Blood 98:913
60. Bose R, Verheij M, Haimovitz-Friedman A, Scotto K, Fuks Z, Kolesnick R (1995) Cell 82:405
61. Verheij M, Bose R, Lin XH, Yao B, Jarvis WD, Grant S, Birrer MJ, Szabo E, Zon LI, Kyriakis JM, Haimovitz-Friedman A, Fuks Z, Kolesnick RN (1996) Nature 380:75
62. Kaufmann SH (1998) Biochim Biophys Acta 1400:195
63. Green PS, Leeuwenburgh C (2002) Biochim Biophys Acta 1588:94
64. Clementi ME, Giardina B, Di Stasio E, Mordente A, Misiti F (2003) Anticancer Res 23:2445
65. Cole SP, Bhardwaj G, Gerlach JH, Mackie JE, Grant CE, Almquist KC, Stewart AJ, Kurz EU, Duncan AM, Deeley RG (1992) Science 258:1650
66. Allen JD, Jackson SC, Schinkel AH (2002) Cancer Res 62:2294
67. Lehnert M (1996) Eur J Cancer 32A:912
68. Binaschi M, Supino R, Gambetta RA, Giaccone G, Prosperi E, Capranico G, Cataldo I, Zunino F (1995) Int J Cancer 62:84
69. van Hattum AH, Pinedo HM, Schluper HM, Erkelens CA, Tohgo A, Boven E (2002) Biochem Pharmacol 64:1267
70. Friesen C, Fulda S, Debatin KM (1999) Cell Death Differ 6:471
71. Choi CH, Kim HS, Kweon OS, Lee TB, You HJ, Rha HS, Jeong JH, Lim DY, Min YD, Kim MS, Chung MH (2000) Mol Cells 10:38

72. Kuninaka S, Ichinose Y, Koja K, Toh Y (2000) Br J Cancer 83:928
73. Nielsen D, Maare C, Skovsgaard T (1996) Gen Pharmacol 27:251
74. Zunino F, Gambetta RA, Zaccara A, Carsana R (1981) Tumori 67:399
75. Osbild S, Brault L, Battaglia E, Bargel D (2006) Anticancer Res 26:3595
76. Pompella A, Corti A, Paolicchi A, Giommarelli C, Zunino F (2007) Curr Opin Pharma-col 7:360
77. Ravi D, Das KC (2004) Cancer Chemother Pharmacol 54:449

Top Curr Chem (2008) 283: 21–44
DOI 10.1007/128_2007_11
© Springer-Verlag Berlin Heidelberg
Published online: 15 November 2007

Anthracycline Cardiotoxicity

Pierantonio Menna[1,2] · Emanuela Salvatorelli[1,2] · Luca Gianni[3] ·
Giorgio Minotti[1,2] (✉)

[1]University Campus Bio-Medico of Rome, CIR and Drug Sciences,
Via Alvaro del Portillo 21, 00128 Rome, Italy
g.minotti@unicampus.it

[2]Department of Drug Sciences and Center of Excellence on Aging,
G. d'Annunzio University School of Medicine, Via dei Vestini, 66013 Chieti, Italy

[3]Division of Medical Oncology A, Istituto Nazionale per lo Studio e la Cura dei Tumori,
Via Venezian 1, 20133 Milan, Italy

Abstract The clinical use of doxorubicin and other quinone-hydroquinone anticancer anthracyclines is limited by a dose-related cardiotoxicity. Here, we review the correlation of cardiotoxicity of doxorubicin with its peak plasma concentration and diffusion in the heart, followed by reductive bioactivation or oxidative inactivation. One-electron quinone reduction and two-electron side chain carbonyl reduction are accompanied by iron and free radical reactions that are responsible for many aspects of anthracycline cardiotoxicity. In contrast, one-electron hydroquinone oxidation serves as a salvage pathway for degrading and detoxifying anthracyclines. Mechanism-based cardioprotective strategies therefore involve replacing bolus administration with slow infusions (to reduce the drug's plasma peak), encapsulating anthracyclines in liposomes (to reduce the drug's cardiac diffusion), and administering antioxidants or iron chelators. Preclinical modelling and clinical studies suggest that eliminating the side chain carbonyl group reduction warranted a satisfactory degree of cardioprotection. Approved or investigational anthracyclines that lacked the carbonyl group or showed an inherent resistance to carbonyl reduction might prove safer than doxorubicin, particularly when administered with new generation drugs that otherwise caused a toxic synergism with doxorubicin.

Keywords Analogs · Anthracyclines · Cardiotoxicity · Metabolism · Pharmacokinetics

Abbreviations

DOX(OL)	doxorubicin(ol)
EPI(OL)	epirubicinol
DNR(OL)	daunorubicin(ol)
IDA(OL)	idarubicin(ol)
CHF	congestive heart failure
Pgp	P glycoprotein
C_{max}	peak plasma concentration
AUC	area under the curve
RyR2	ryanodin receptor-2
$O_2{}^{\cdot-}$	superoxide anion
H_2O_2	hydrogen peroxide
ROS	reactive oxygen species
$\cdot$OH	hydroxyl radical
IRP-1	iron regulatory protein 1
$Mb^{II}O_2$	oxyferrous myoglobin
$Mb^{IV=O}$	ferrylmyoglobin
Mb^{III}	metmyoglobin
LVEF	left ventricular ejection fraction
DIDOX	C-13 deoxy-5 imino doxorubicin
PTX	paclitaxel
DCT	docetaxel
VEGF	vascular endothelial growth factor

1
General Concepts

In spite of their longer than 40 years record of longevity, the anthracyclines remain among the most effective cytotoxics available to the oncologist. They act primarily by forming a stable ternary complex with DNA and topoisomerase II [1]. The skeleton of anthracycline is composed of a tetracyclic ring system with adjacent quinone-hydroquinone moieties, an aminosugar (daunosamine) bound by a glycosidic bond to C-7, and a short side chain with a carbonyl group at C-13. Relatively minor changes in this skeleton result in major changes of the spectrum of antitumor activity. Anthracyclines with a primary alcohol at the side chain terminus, like doxorubicin (DOX) and epirubicin (EPI), show activity against both solid and hematologic malignancies. Anthracyclines with a methyl group, like daunorubicin (DNR) and idarubicin (IDA), are used primarily to treat acute myeloblastic leukaemia and AIDS-related Kaposi's sarcoma. EPI differs from DOX in a positional

Table 1 Anthracycline structure and effects of substituents on the clinical spectrum of antitumor activity

ANTHRACYCLINE	R_1	R_2	R_3	APPROVED INDICATIONS
DOX	CH_2OH	OCH_3	OH ax	Carcinomas, sarcomas,
EPI	CH_2OH	OCH_3	OH eq	lymphomas
DNR	CH_3	OCH_3	OH ax	Acute myeloblastic leukaemia,
IDA	CH_3	H	OH ax	AIDS-relatedKaposi's sarcoma

change of the hydroxyl group at C-'4 in daunosamine, while IDA differs from DNR in the absence of a methoxy substituent at C-4 (Table 1).

Anthracyclines have long been known to induce also cardiotoxicity, an untoward effect that limits their clinical use. Patients exposed to anthracyclines may experience arrhythmias, hypotension, and mild depression of myocardial contractility; in a few cases, myocarditis and pericardial effusions may also occur. This is the so-called *acute cardiotoxicity*, a reversible and usually benign condition that develops shortly after one or two doses of an anthracycline. Acute cardiotoxicity is relatively infrequent (~1% of patients), and does not represent an indication to interrupt an anthracycline-based regimen [1, 2]. Unfortunately, however, anthracyclines may also cause dilative cardiomyopathy and congestive heart failure (CHF). In the absence of individual risk factors (age, hypertension, preexisting arrhythmias or valvular diseases, diabetes or other metabolic disturbances) cardiomyopathy and CHF only develop when the "lifetime" cumulative dose of DOX exceeds a threshold that is currently set at ~450 mg/m^2 [3]. "Lifetime" means that the cumulative dose of DOX should always be intended as the arithmetic sum of the individual doses administered to a patient, even if months or years occurred between one cycle and the other. This concept rests with the fact that the cardiac half-life of DOX may be surprisingly long, such that anthracycline-related material could be measured in the heart of patients deceased long after the last administration of DOX [4]. Cardiomyopathy therefore develops in response to the summation of injuries inflicted by de novo DOX and long-lived cardiac residues of prior DOX.

Retrospective clinical studies have also shown that the severity of cardiac damage in a given patient is inversely correlated to the levels of P glyco-protein (Pgp) in the endothelium of arterioles and capillaries of the heart of that patient [5]. Pgp is a prominent member of the superfamily of ATP-binding proteins that extrude anthracyclines and many other drugs from within the cells into the extracellular fluids. Thus, the dose-dependence of cardiomyopathy and its inverse correlation with Pgp highlight a close link be-tween myocardial damage and the amount of DOX that accumulates in the heart.

Anthracycline-induced cardiomyopathy and CHF exhibit a time-related pat-tern. In the vast majority of patients these cardiac events occur within a year from the completion of a cumulative anthracycline regimen [2], but very late forms of cardiomyopathy are not uncommon among the long-term survivors of childhood cancer [6]. Cardiomyopathy and CHF therefore represent what is popularly referred to as *chronic cardiotoxicity*. Full-blown chronic cardiotoxic-ity is a life-threatening condition, especially if one appreciates that it may show only partial or transient responsiveness to cardiovascular drugs like digitalis, angiotensin converting enzyme inhibitors, and β-blockers [7].

2
Pharmacokinetic Determinants of Cardiotoxicity

The relation between the myocardial content of DOX and the development of cardiomyopathy anticipates that the risk of cardiac events will depend on a few established pharmacokinetic determinants. Thus, a retrospective analysis of several clinical studies demonstrates that the development of chronic car-diomyopathy correlates with the peak plasma concentration of DOX (C_{max}); in laboratory animals, this correlates with the post-administration ventricular peak of DOX and of its major metabolites [8]. This having being said, one can understand how chronic cardiotoxicity occurs more frequently when a patient is given DOX by i.v. bolus over 5–15 min, a modality that generates high C_{max} values; conversely, chronic cardiotoxicity occurs less frequently (or at higher anthracycline cumulative doses) when DOX is administered by continuous in-fusion over 2 or 4 h, a modality that generates a much lower C_{max} while also producing equal or higher Area Under the Curve (AUC) and objective tumor response [9]. Note, however, that the safety of replacing bolus administration with slow infusions was observed in studies of adult patients but not in pediatric settings, as if the pharmacokinetic determinant(s) of cardiotoxicity depended on age-related factors that await further clarification [10].

Further evidence that the cardiotoxicity of DOX correlates with its C_{max} and diffusion into the heart comes from the successful strategy of encapsulating the anthracycline molecule in liposomal delivery systems. Liposomal DOX reaches high C_{max} values and diffuses through the discontinuous "leaky" endothelium of tumors; however, the liposomes are too big to diffuse through the normal microvasculature of the heart. Thus, the pharmacokinetic harm associated with a high C_{max} of DOX is counterbalanced by a limited partitioning of DOX into the vulnerable cardiomyocytes.

Two liposomal formulations of DOX have been developed and approved for clinical use: an uncoated formulation (currently marketed as Myocet®) and a sterically stabilized polyethyleneglycol-coated formulation (Caelyx®). An uncoated liposomal formulation of daunorubicin is also available (DaunoXome®).

Myocet® has been examined in two phase III studies for the treatment of metastatic breast cancer [11, 12]. In both studies Myocet® demonstrated a mg for mg equipotency to free DOX, but it showed a dramatically reduced cardiac toxicity. Myocet® was also probed in metastatic breast cancer patients who had received DOX in the adjuvant setting. Also in these patients Myocet® showed a good activity vis-à-vis a limited cardiotoxicity, a finding that conceptually re-opened the door to the use of DOX in patients with a previous exposure to anthracyclines [13]. Caelyx® was designed to escape degradation by cells of the reticuloendothelial system and hence, to generate a blood circulation time and an intratumoral drug-liposomal deposition even higher than those obtained with Myocet®. In agreement with these premises Caelyx® showed very high activity but limited cardiotoxicity in patients with breast or ovary cancer; unfortunately, however, the unique pharmacokinetic effects of the polyethylenglycol coating also resulted in a dose-limiting hand-foot syndrome [14].

DaunoXome® has been approved by the Food and Drug Administration as first line therapy of AIDS-related Kaposi's sarcoma; it shows activity and tolerability as a single agent or in combination with other drugs also in refractory/relapsed acute myeloblastic leukaemia, recently diagnosed or recurrent/refractory multiple myeloma, and poor-prognosis non-Hodgkin lymphoma [15, 16]. In all such settings DaunoXome® exhibits an encouraging cardiac tolerability. Of note, limited phase I dose-escalating studies indicate that DaunoXome® may be active and cardiac-safe also in metastatic breast cancer, a clinical setting in which DNR lacked a formal indication [17]. These findings denote the many favorable effects of liposomal encapsulation on the therapeutic index and spectrum of activity of DNR.

3
Metabolic Determinants of Cardiotoxicity

In cardiomyocytes anthracyclines cause vacuolar degeneration, mitochondrial inclusions, myofibrillar disarray and dropout, increased number of lysosomes, apoptosis, and necrosis [1]. This morphologic pattern is seen almost universally in patients and laboratory animals, but the underlying molecular mechanisms and/or chemical mediators may be species-specific.

DOX per se was shown to downregulate cardiac-specific transcriptional regulatory proteins, causing a reduced expression of the Ca^{2+}-gated Ca^{2+} release channel (ryanodin receptor-2, RyR2), α-actin, myosin light chain 2 slow, and others [18–20]. Doxorubicin per se was also shown to bind to cardiolipin, highly abundant in heart mitochondria, thereby altering the normal assembly and functional coupling of the respiratory chain [21]. Whether such mechanisms contributed to inducing full-blown cardiotoxicity remained uncertain, especially in the light of the rather high doses of DOX that had to be used in vitro or in animal models; therefore, the current thinking is that DOX gained more toxicity upon conversion to reactive metabolites or intermediates.

3.1
One-Electron Reductive Bioactivation

The bioactivation of anthracyclines to cardiotoxic species offers a typical example of structure-activity relations. One-electron reduction of the quinone moiety of DOX results in the formation of a semiquinone free radical which regenerates its parent quinone by reducing molecular oxygen to superoxide anion ($O_2^{\cdot-}$) and hydrogen peroxide (H_2O_2), members of the broad family of reactive oxygen species (ROS) (Fig. 1A).

Such a "redox-cycling" of DOX is supported by NADPH-dependent cytochrome P450 or b_5 reductases, mitochondrial NADH dehydrogenases (NADH ubiquinone oxidoreductase and the so-called "exogenous reductase"), xanthine dehydrogenase, and endothelial nitric oxide synthase (reductase domain). One-electron reduction of DOX is accompanied also by a reductive release of Fe(II) from ferritin. On the one hand, the anthracycline semiquinone ($E^{o'} =- 0.4$ V) redox-couples with the transprotein channels of ferritin and initiates an electron tunnelling that reduces and mobilizes the polynuclear ferric oxohydroxide core of ferritin ($E^{o'} =- 0.23$ V); on the other hand, a semiquinone-derived $O_2^{\cdot-}$ would be small and electronegative enough ($E^{o'} =- 0.33$ V) to penetrate the transprotein channels and to reduce ferritin iron directly [22]. All such processes elevate the cellular levels of Fe(II) ions that convert the $O_2^{\cdot-}$ and H_2O_2 into hydroxyl radical ($^{\cdot}$OH), one of the most potent oxidants possibly formed in biologic systems ($E^{o'} = 2.31$ V at pH 7.0) [23]. A one-electron reduction of DOX and other anthracyclines would therefore be expected to cause oxidative stress, especially if one appreciates that cardiomyocytes are constitutively ill-equipped with $O_2^{\cdot-}$ or H_2O_2-detoxifying enzymes like superoxide dismutase or catalase; in addition, DOX-derived H_2O_2 quickly inactivates selenium-dependent glutathione peroxidase and decreases the protein levels and activity of cytosolic copper- and zinc-dependent superoxide dismutase [24, 25].

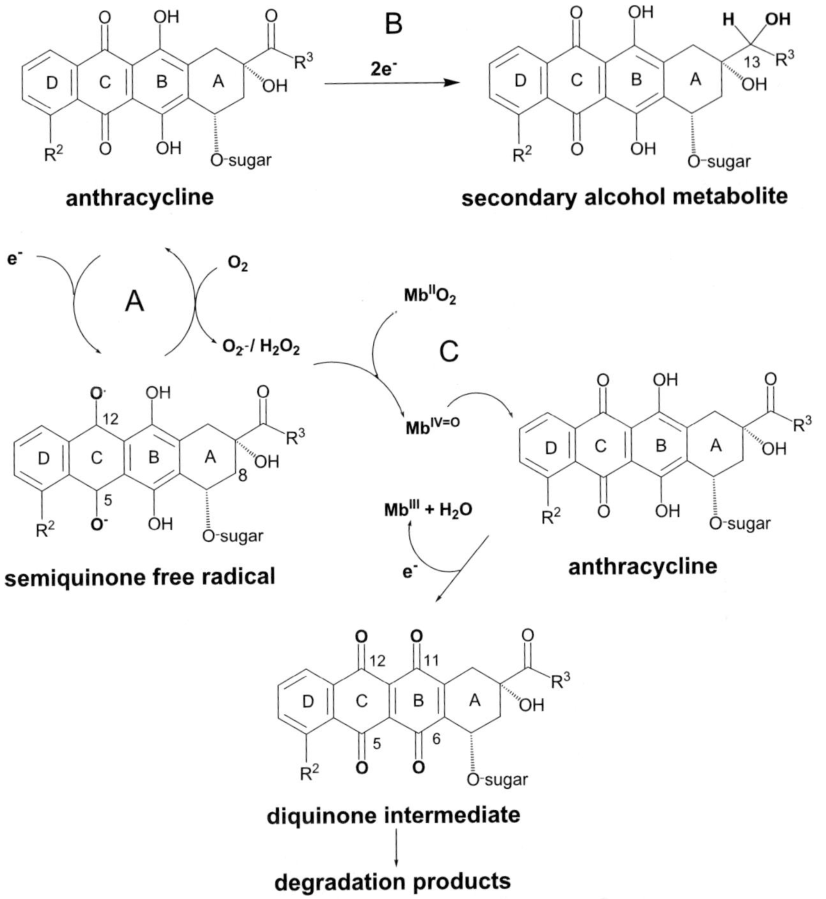

Fig. 1 Anthracycline reductive bioactivation vs. oxidative degradation and detoxification. **A** One-electron quinone reduction; **B** Two-electron side chain carbonyl reduction; **C** One electron hydroquinone oxidation. $O_2^{\cdot-}$, superoxide anion; H_2O_2, hydrogen peroxide; $Mb^{II}O_2$, oxyferrous myoglobin; $Mb^{IV=O}$, ferrylmyoglobin

3.2
Two-Electron Reductive Bioactivation

Anthracyclines are said to gain toxicity also after a two-electron reduction of the side chain C-13 carbonyl moiety. In laboratory animals such reaction is mediated by heterogeneous families of cytoplasmic NADPH-dependent aldo/keto- or carbonyl-reductases, but in human myocardial samples it seems to be mediated almost exclusively by a specific family of aldehyde reductases [26, 27]. Two-electron carbonyl reduction converts anthracyclines to secondary alcohol metabolites referred to as doxorubicinol (DOXOL), epirubicinol (EPIOL), daunorubicinol (DNROL), or idarubicinol

(IDAOL) (Fig. 1B). In cell-free systems or isolated rat heart preparations these metabolites may be ~30–40 times more potent than their parent anthracyclines at inactivating ATP-dependent Ca^{2+}-handling proteins [28–30]; DOXOL is also more potent than DOX at suppressing the gene expression of RyR2 [31]. Perhaps more importantly, secondary alcohol metabolites exhibited a unique reactivity toward the Fe-S cluster of cytoplasmic aconitase/Iron Regulatory Protein-1, an important and versatile regulator of iron homeostasis, energy metabolism, and redox balance of the cell [22, 32, 33].

While attesting to a unique biochemical reactivity of secondary alcohol metabolites as compared with their parent drugs, the pharmacodynamics of such metabolites and their possible contribution to cardiomyopathy and CHF

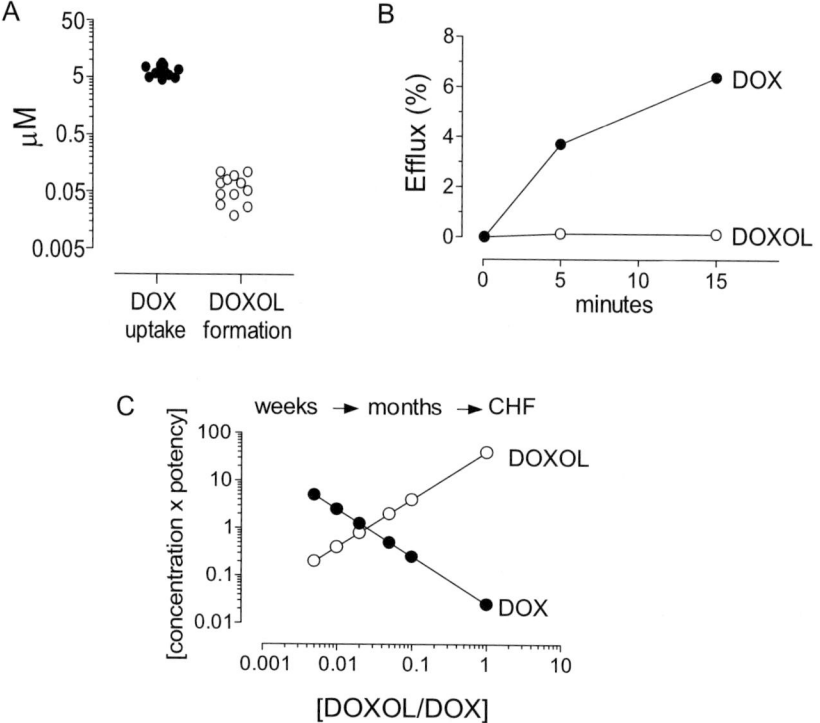

Fig. 2 Pharmacokinetics and toxicokinetics of DOX and DOXOL in human myocardium. **A** DOX uptake and DOXOL formation were measured in human myocardial strips incubated in plasma with 10 μM DOX for 4 h. **B** Human myocardial strips were incubated in plasma with 10 μM DOX for 4 h and then placed in anthracycline-free plasma which was assayed for DOX and DOXOL efflux. The values were expressed by normalizing DOX(OL) in plasma to DOX(OL) in the strips. **C** DOXOL/DOX ratios were plotted by assuming an immediate post-administration concentration of 10 or 0.05 μM for DOX or DOXOL, respectively (cfr. *panel A*). [Concentration × potency] factors were calculated by assuming that DOXOL was 40-times more potent than DOX toward ATP-dependent Ca^{2+} handling proteins; based on data in [26, 27, 29]

should be interpreted with due caution. In human myocardial samples the fractional conversion of DOX and EPI to DOXOL or EPIOL usually averaged $\leq$1% [26, 27, 34, 35]: this would be too low for DOXOL or EPIOL to play a major role in cellular damage, even if one considered that the metabolites were ~30–40 times more reactive than DOX or EPI.

The role of secondary alcohol metabolites must therefore be appraised also on pharmacokinetic grounds. Unmodified anthracyclines are lipophilic enough to diffuse from cardiomyocytes back into plasma, while secondary alcohol metabolites are too polar to do so and show essentially no efflux [27] (see Fig. 2A,B for a comparison of DOX with DOXOL). It is because of such differences that the post-administration cardiac levels of DOX gradually decreased by $\geq$2 orders of magnitude while those of DOXOL remained stationary or increased up to a level where DOXOL eventually equalled DOX [4]. One can therefore approximate that the [potency x concentration] factor of DOX only transiently prevailed after DOX administration, while that of DOXOL increased and eventually prevailed at the time when cardiomyopathy and CHF became clinically evident (Fig. 2C).

Several lines of evidence support the notion that secondary alcohol metabolites might represent the long-lived anthracycline species that best correlates with a development of cardiomyopathy and CHF: (i) in laboratory animals the development of chronic cardiomyopathy correlated with a time-related accumulation of DOXOL in the heart [28], (ii) mice with a cardiac-specific overexpression of anthracycline carbonyl reductases exhibited an increased conversion of DOX to DOXOL and an accelerated course of development of chronic cardiomyopathy [36], (iii) mice with the genetic deletion of one copy of the carbonyl reductase allele formed fewer amounts of DOXOL and developed a less severe cardiotoxicity [37].

3.3
One-Electron Oxidative Degradation and Detoxification

For many years it was thought that one- or two-electron additions would be the only metabolic pathways that influenced anthracycline toxicity in the heart. Over the last few years it turned out that anthracyclines are liable also to one-electron oxidation, a process that was brought into focus through an appreciation of the dual role of myoglobin in the heart. On the one hand, oxyferrous myoglobin ($Mb^{II}O_2$) plays an established role in storing and exchanging oxygen with mitochondria or other cellular sites that need a steady-state oxygen supply; on the other hand, an oxidation of $Mb^{II}O_2$ with H_2O_2 generates a long-lived iron-oxo moiety ($Mb^{IV=O}$) that is formally identical to the compound II of peroxidases and catalyzes the oxidation of many biomolecules [38]. Studies with cell-free systems uncovered that the redox cycling of DOX, and the consequent formation of H_2O_2, could very well be accompanied by $Mb^{IV=O}$ formation and one-electron oxidation of the B ring hydroquinone in juxtaposition to the C ring quinone. During the course of this reaction $Mb^{IV=O}$ reduced to metmyoglobin (Mb^{III}), while DOX lost its fluorescent and chromatographic properties in a manner that was highly suggestive of anthracycline ring opening and degradation [39].

The precise chemistry of anthracycline oxidative degradation is only partially understood. An $Mb^{IV=O}$-dependent oxidation was observed also with DNR and IDA or a simple quinone-hydroquinone compound like napht-hazarin, but a less extensive oxidation occurred with an anthracycline analog (aclarubicin) that lacked the hydroquinone moiety [39]. In confirming that oxidation occurred at the hydroquinone moiety, these results suggested that ring opening might occur through the formation of a highly unstable diquinone intermediate (Fig. 1C) [39, 40]. 3-Methoxyphthalic acid, an oxidatively modified remnant of the D ring, was shown to be formed in vitro during oxidation of DOX with the compound I of authentic peroxidases or H_2O_2/Mb^{III} mixtures that also formed a compound I-like species with a porphyrin π cation radical [41, 42]; however, 3-methoxhyphthalic was not detected during oxidation of DOX with H_2O_2 and the physiologic $Mb^{II}O_2$, which only generates a compound II [43]. Regardless of such mechanistic uncertainties, in vitro studies offered unambiguous evidence that neither purified 3-methoxyphthalic acid, nor DOX samples that had been oxidized with $H_2O_2/Mb^{II}O_2$, were able to damage cardiomyocytes [41, 43].

The aforesaid observations suggested that one-electron oxidation might serve to detoxify DOX in competition with its bioactivation by one- or two-electron reduction, but one such concept had to be probed in a cellular environment in which anthracycline reduction and oxidation occurred simultaneously. This experimental approach became possible upon discovering that *tert*-butoxycarbonyl-alanine introduced sterical barriers to a reaction of DOX with $Mb^{IV=O}$, thereby blocking DOX degradation. When administered to cardiomyocytes *tert*-butoxycarbonyl-alanine increased the steady state levels of DOX, its conversion to ROS, and its concentration-related toxicity [43]. On balance, these results support the notion that anthracycline oxidation operates a salvage mechanism against the reductive bioactivation of DOX in cardiomyocytes.

4
Translating the Metabolic Determinants of Cardiotoxicity into Protective Strategies

4.1
Antioxidants

The possible role of ROS in anthracycline-induced cardiotoxicity offered a rationale for using antioxidants that intercepted ROS and/or increased the ROS resistance of vulnerable targets like membrane lipids or labile thiols. In pre-clinical models, good results were obtained with such diverse compounds as probucol, piperidine nitroxides, spin traps, melatonin, vitamin E, membrane-permeable thiol-reducing agents (like N-acetylcysteine), and rutoside-type flavonoids [1]. In clinical settings, similar strategies were largely unsuccess-ful: for example, robust doses of vitamin E or N-acetylcysteine neither de-layed nor mitigated cardiotoxicity induced by cumulative doses of DOX [44]. Carvedilol, a β-blocker with potent but only partially characterized antiox-idant and antiapoptotic properties [45], showed a promising protective ef-ficacy in patients treated with DOX [46]; however, the study design and statistical power were questioned [47]. The protective efficacy of carvedilol should be probed in randomized trials that used nonantioxidant β-blockers in the control arm.

4.2
Iron Chelators

Another protective strategy rests with the chelation of iron by dexrazoxane, a bis-ketopiperazine which hydrolyzes to a diacid-diamide structurally simi-lar to EDTA. The hydrolysis of dexrazoxane is a complex and fascinating process: it may occur enzymatically through the action of dihydroorotases, but it may also occur nonenzymatically upon ligand binding interactions with metals [48]. Dexrazoxane shows an extraordinarily rapid diffusion in car-diomyocytes, and meets the structural requirements to chelate iron before it catalyzed the conversion of $O_2{}^{\cdot-}$ and H_2O_2 into $^{\cdot}OH$. Dexrazoxane prevented histologic lesions and contractile dysfunction induced by DOX in both pre-clinical and clinical studies, granting an FDA approval for its use in patients who were scheduled to continue on DOX or EPI after a prior exposure to a cumulative dose of 300 mg of DOX/m^2. [1, 49]. Here it is worth noting that dexrazoxane had been suspected to reduce response rates in women receiv-ing DOX for the treatment of metastatic breast cancer [50], but the current thinking is that such an effect may have been overestimated [51].

In evaluating the lack of clinical protection by antioxidants vis-à-vis the protection afforded by dexrazoxane one cannot escape the conclusion that the role of iron may well extend beyond $^{\cdot}OH$ formation. In the light of the avid-ity of secondary alcohol metabolites for aconitase/IRP-1 one might envision that anthracyclines disrupt iron homeostasis in the heart, thereby causing the misplacement of iron ions at cellular sites that lose their function after their sterical occupation by this metal: this might very well be the case of the RyR2, which undergoes inactivation upon coordination of iron by the cys-teine residues lining the channel [52]. Thus, dexrazoxane might protect the heart by preventing iron misplacement rather than iron-catalyzed free radical reactions. Similar conclusions were reached during the preclinical testing of other iron chelators, like e.g., lipophilic aroylhydrazones [53, 54].

4.3
Noncardiotoxic Anthracycline Analogs

The promising results obtained with slow infusions or liposomal anthracyclines did not halt the search for analogs that caused less cardiac toxicity while also showing antitumor activity. There are several reasons to explain the unabated search for such a "better anthracycline". Liposomal anthracyclines are quite expensive, and many doctors perceive them as too laborious to prepare and/or to infuse; moreover, as mentioned, pegylated liposomal DOX causes a severe dose-limiting hand-foot syndrome [14]. Slow infusions too are felt as a last-resort and laborious procedure, not to mention their uncertain value in pediatric settings.

The search for a "better anthracycline" has produced some 2000 analogs, but only a handful of them reached the stage of consideration for clinical use. Before reviewing the main features of some newly designed anthracyclines, it may be worth considering that the currently approved "old" anthracyclines do exhibit discrete differences in their cardiotoxic potential. There have been variable levels of interest in examining such differences; here, we will briefly review some lessons from IDA or EPI.

Idarubicin (administered by the canonical iv route) was shown to decrease the left ventricular ejection fraction (LVEF) in anthracycline-naïve patients, or to cause CHF in patients with preexisting cardiovascular disease or prior anthracycline treatment [55]. In the light of its very high lipophilicity, IDA was subsequently administered also per os, and the studies that adopted this route showed essentially no cardiotoxicity [56]. In examining the different outcomes of these studies one may want to consider that an equiactive plasma drug exposure would be achieved following oral doses of IDA $\sim$2.5-fold higher than the i.v. doses [57]. This was not the case in the available studies, which usually adopted less than bioequivalent doses of IDA; hence, the cardiac tolerability of oral IDA might well be a reflection of its lower bioavailability. Whether IDA displayed a diminished cardiotoxicity awaits further clarification.

A more instructive lesson is offered by EPI, which only differs from DOX in an axial-to-equatorial epimerization of the hydroxyl group at $C-'4$ in daunosamine (cfr. Table 1). This positional change has long been known to facilitate the glucuronidation and body clearance of EPI as compared to DOX; therefore, the dose of EPI equiactive to 1 mg of DOX increases to 1.5 mg [58]. Interestingly, however, the dose of EPI equicardiotoxic to 1 mg of DOX may be as high as 1.8–2 mg, as if glucuronidation and body clearance were not the only factor that diminished the dose-related cardiotoxicity of EPI [59]. In practice, cardiomyopathy and CHF would not develop until the cumulative dose of EPI exceeded 850–900 mg/m^2, a dose level at which EPI would be $\sim$1.3 times more active than DOX.

Recently, we demonstrated that EPI might cause less cardiotoxicity than DOX because of its reduced intramyocardial conversion to ROS or its secondary alcohol metabolite EPIOL. In human myocardial samples the bioactivation of EPI to ROS was limited by a unique mechanism of anthracycline protonation-sequestration in acidic organelles like recycling endosomes, lysosomes, or vesicles of the *trans*-Golgi network; this resulted in a limited partitioning of EPI toward the mitochondrial sites of one-electron redox cycling [60]. The avidity of EPI for the acidic organelles could not be attributed to a commensurate avidity for protons, as the protonatable aminogroups of DOX and EPI shared similar pK_a values (8.34 vs. 8.08); instead, it was caused by the higher lipophilicity of EPI compared to DOX (1.1 vs. 0.5, as determined by octanol:Tris buffer partitioning) [61], with this factor increasing the net amount of EPI that partitioned into the vesicles and became available to protonation-sequestration [60]. While exhibiting very little or no biotransformation to ROS, EPI also formed ~50% less alcohol metabolite than DOX. This was caused primarily by an impaired catalytic specificity of EPI for the same aldehyde reductases that converted DOX to DOXOL [27, 60]. It is worth emphasizing that neither a protonation-sequestration mechanism nor a defective conversion to EPIOL would be expected to diminish the activity of EPI in tumor cells. Drug-naïve tumor cells usually show a defective acidification of their vesicular apparatus, and consequently fail to sequester anthracyclines [62]; furthermore, secondary alcohol metabolites do not always mediate or sometimes diminish the antitumor activity of anthracyclines, although the molecular foundations of such a reduced activity have not been formally elucidated [1].

In providing formal explanations about how EPI caused cardiomyopathy and CHF at doses higher than equiactive to DOX, the aforesaid results clearly illustrate that anthracyclines lacking the formation of ROS or secondary alcohol metabolites may fulfill—at least in principle—the characteristics of good activity vis-à-vis reduced toxicity. One should also note that, in the case of EPI, a ~30% reduction of its dose-related cardiotoxicity would correlate much better with a ~50% reduction of EPIOL formation rather than with a near-to-complete abrogation of ROS formation. This might reinforce the concept that secondary alcohol metabolites were more important than ROS in promoting the development and/or progression of cardiotoxicity.

As mentioned, very many "new" anthracyclines have been designed in an attempt to overcome the problem of cardiotoxicity. Some of them should not be viewed as "analogs" obtained by chemical modifications of the anthracycline chromophore and/or aminosugar; instead, they represent prodrugs that deliver authentic DOX to tumor cells but not to cardiomyocytes, similar to what is obtained by entrapping DOX in liposomes. A good example is offered by DOX covalently linked to *N*-glutaryl-[4-hydroxyprolyl]-Ala-Ser-cyclohexaglycyl-Glu-Ser-Leu. This prodrug (code-named L-377,202) releases DOX or diffusible leucine-DOX in prostate cancer cells after a cleavage of its peptide moiety by the serine protease activity of the prostate cancer specific antigen [63].

Other prodrugs are obtained by linking DOX to polymers recognized by tumor-specific receptors. A good example of this second strategy is offered by PK2 (DOX linked to hydroxypropyl-methacrylamide-galactosamine copolymers recognized by the liver-specific asialoglycoprotein receptor). PK2 holds promise for an effective and safe therapy of primary hepatocarcinomas [64].

The number of true "analogs" keeps growing year after year. In returning to comparisons of DOX with EPI, we will focus on those analogues that proved to form fewer amounts of ROS and/or secondary alcohol metabolite; this will serve as an opportunity to verify whether the free radical or alcohol metabolite hypotheses of cardiotoxicity translated into chemical entities more advantageous than EPI.

Attempts to eliminate an involvement of DOXOL in cardiotoxicity formed the rationale to design C-13 deoxydoxorubicin (Fig. 3).

When probed in a rabbit chronic model of cardiotoxicity C-13 deoxydoxorubicin caused essentially no effect on major indices of myocardial contractility nor did it suppress the expression of RyR2 [31]. The next logical step was to design an anthracycline (5-imino, C-13 deoxydoxorubicin, provisionally referred to as DIDOX), in which also the quinone moiety had been modified to eliminate the formation of ROS (Fig. 3) [65]. DIDOX caused less cardiotoxicity than DOX in laboratory animals, but there is little or no room to conclude that it caused less cardiotoxicity than C-13 deoxydoxorubicin. If anything, it was noted that DIDOX caused ~3–4 times less myelotoxicity than DOX. Inasmuch as myelotoxicity is a surrogate of the antiproliferative activity of any given chemotherapeutic, these findings anticipate that an elimination of the quinone moiety diminished the antitumor effects of DIDOX. In principle, this might be caused by perturbances of the formation of an anthracycline-topoisomerase II-DNA complex, which relies on a precise overlap of the B and C rings with adjacent base pairs of DNA; it might also reflect the elimination of ROS-dependent factors like telomere oxidation, formation and adduction of malondialdehyde to DNA, formation of anthracycline-formaldehyde conjugates with longer cellular half-life and higher DNA cross-linking activity [1].

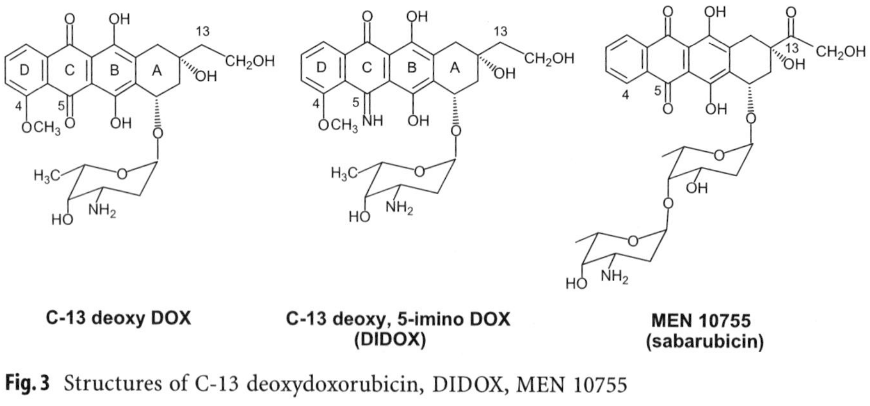

C-13 deoxy DOX   **C-13 deoxy, 5-imino DOX**   **MEN 10755**
                                    (DIDOX)   **(sabarubicin)**

Fig. 3 Structures of C-13 deoxydoxorubicin, DIDOX, MEN 10755

Face-to-face comparisons of C-13 deoxydoxorubicin with DIDOX suggest that an elimination of alcohol metabolite formation suffices at reducing cardiotoxicity while not impairing antitumor activity. On the basis of what we know from EPI, it also seems that changes in alcohol metabolite formation do not always require chemical modifications at C-13 but may well occur after introducing modifications at a distance from C-13.

The most persuasive validation of these concepts comes from preclinical and clinical studies of sabarubicin, formerly code-named MEN 10755. Sabarubicin is the lead compound of disaccharide anthracyclines that were designed to explore the effects of the aminogroup of daunosamine on the inhibition of topoisomerase II: it is obtained by removal of the methoxy substituent at C-4 in the aglycone and by insertion of 2,6-dideoxy-L-fucose between the aglycone and daunosamine (Fig. 3). Studies with several tumor cell lines and human tumor xenografts show that sabarubicin is equiactive or sometimes superior to DOX [66]; it may show activity also in cell lines that developed resistance to DOX because of the overexpression of antiapoptotic factors like Bcl-2 [67]. Early preclinical trials showed that the higher antitumor activity of sabarubicin occurred in the face of its defective conversion to sabarubicinol, which was of the same order of magnitude as that described for the defective conversion of EPI to EPIOL; however, sabarubicinol—but not DOXOL or EPIOL—showed also a limited reactivity toward the Fe – S cluster of aconitase/IRP-1 [68, 69]. From a structure-activity view point it was shown that the formation of sabarubicinol was limited by both the lack of the methoxy group at C-4 and the presence of the disaccharide moiety, while the [Fe – S]-reactivity of sabarubicinol was only limited by the presence of the disaccharide moiety [68]. Because of such unique properties, studies with rats proved that sabarubicin was appreciably less cardiotoxic than equimyelotoxic DOX or EPI [70].

The lessons from sabarubicin are twofold. On the one hand, the studies with sabarubicin confirm that modifications at distance from the side chain can diminish carbonyl reduction and improve the cardiac safety of an anthracycline while also sparing, if not improving its antitumor activity. In extending this concept to the pharmaceutical engineering of anthracyclines one would expect that combining modifications of the number of sugar moieties with epimerization of the distal daunosamine might improve the therapeutic index of an anthracycline even further. On the other hand, sabarubicin offered a sound opportunity to translate the alcohol metabolite hypothesis of cardiotoxicity into clinical trials. In a phase I study of 24 patients with advanced solid tumors only two patients experienced an asymptomatic decrease of the LVEF [71], while in a phase II study of patients with advanced platinum/taxane-resistant ovarian cancer no signs or symptoms of CHF were observed [72]. And finally, a phase II study of patients with progressive hormone refractory prostate cancer showed that moderate-to-severe cardiotoxicity only occurred in 3 of 32 patients exposed to cumulative doses of >500 mg of sabarubicin/m^2 [73]. It is hoped that such promising results will soon be confirmed in randomized phase III clinical trials.

4.4
Modulators of Anthracycline Degradation

Drugs that stimulated anthracycline oxidation might prove useful to diminish the cardiac levels of DOX and its reductive bioactivation to DOXOL or ROS. Salicylic acid, a commonly used anti-inflammatory and analgesic drug, was shown to form phenoxyl radicals that dimerized to biphenol quinone, an oxidant able to degrade anthracyclines [74]. These limited observations do not imply that salycilic acid improved the therapeutic index of anthracyclines, as similar reactions might well occur also in tumors. Drugs that accelerated the redox turnover of Mb$^{IV=O}$ with DOX or facilitated sterical interactions between the two (an effect opposite to that of *tert*-butoxycarbonyl-alanine) might prove more specific at inducing DOX degradation in the heart.

5
Cardiotoxic Synergism of Anthracyclines with Other Drugs

The cardiotoxicity induced by DOX has long been known to be aggravated by concomitant pharmacological or physical therapies. Cyclophosphamide is suspected to synergize with DOX in the heart, but the mechanism of such a synergism has not been formally elucidated. Cyclophosphamide, as many other alkylating agents, can randomly and dose-independently cause coronary events [75], but whether a latent myocardial ischemia precipitates the damage induced by anthracyclines should require ad hoc investigations. Chest irradiation is another factor that introduces a definite risk of cardiac events. In the clinical practice, however, radiation therapy and cyclophosphamide-anthracycline regimens remain essential components of the treatment protocol of many cases of breast cancer or other malignancies.

The last few years have witnessed novel clinical paradigms of a toxic synergism between anthracyclines and other drugs, which resulted in symptomatic CHF at lower than expected cumulative doses of DOX. The synergism may be caused by traditional cytostatics like taxanes, or by new generation targeted drugs like the anti HER2/neu monoclonal antibody trastuzumab.

5.1
Anthracyclines and Taxanes

Paclitaxel (PTX) was the first approved member of a family of tubulin-stabilizing agents referred to as taxanes. PTX is active in head, neck, breast, and ovary cancers; its most frequent nonhematologic toxicities include peripheral neuropathy and hypersensitivity reactions, with the latter being caused by its vehicle Cremophor EL. PTX alone did not induce clinically relevant cardiotoxicity, with the possible exception of post-infusion arrhythmias also attributable to Cremophor EL; nonetheless, early pivotal trials of bolus DOX immediately followed by PTX in metastatic breast cancer patients showed that the combination caused a unacceptably high incidence of myocardial dysfunction and symptomatic CHF at cumulative doses ≤ 480 mg of DOX/m^2 [76]. Such an unexpected cardiotoxicity of DOX-PTX could be dealt with by reducing the cumulative dose of DOX to 360 mg/m^2 or by separating the two agents by 4 hours or longer [77]. Combining DOX with the other approved taxane docetaxel (DCT) did not show the same toxic synergism as that observed with the DOX-PT doublet, but the largest phase III trial that demonstrated the safety of DOX immediately followed by DCT was biased by a cautionary reduction of the mean cumulative dose of DOX to 378 mg/m^2 [78].

Over the last few years we were able to show that PTX is an allosteric modulator of the cytoplasmic aldehyde reductases that convert DOX to DOXOL in human myocardium; by this mechanism PTX improves the catalytic efficiency (V_{max}/K_m) and causes a net increase of DOXOL formation in the heart [26, 27] (Table 2).

Table 2 Effects of PTX and DCT on DOXOL or EPIOL formation in human myocardium

Anthra-cycline	Taxane	$K_m{}^a$ (μM)	$V_{max}{}^b$ (nmol/mg prot./min)	$V_{max}/K_m{}^c$ ml/(mg prot./min) ($\times 10^{-4}$)	Alcohol metabolite (μM)
DOX	–	81	0.015	1.9	$0.05 \pm 0.01^*$
	PTX	42	0.043	10	0.08 ± 0.01
	DCT	47	0.036	7.7	0.09 ± 0.02
EPI	–	250	0.0025	0.1	0.03 ± 0.003
	PTX	240	0.003	0.13	0.03 ± 0.01
	DCT	230	0.0027	0.11	0.03 ± 0.004

The kinetics of DOXOL or EPIOL formation (a, b, c) were determined in incubations containing isolated human heart cytosol, NADPH, and appropriate amounts of DOX or EPI, with or without 1 μM taxanes. Net values of DOXOL or EPIOL formation were determined in human myocardial strips incubated in plasma with 10 μM anthracyclines for 4 h, with or without 6 μM PTX or DCT formulated in Cremophor EL or polysorbate 80, respectively.
* Indicates $p < 0.05$ for DOXOL vs. EPIOL and $p < 0.05$ for DOXOL vs. DOXOL + PTX or DOXOL + DCT
Adapted from [27]
PTX, paclitaxel; DCT, docetaxel

A kinetic model was developed to show that clinically relevant concentrations of PTX bind with a high affinity ($^1K_m \sim 1.2\,\mu M$) to the regulatory site of aldehyde reductases, inducing conformational changes that improve both the orientation of DOX in the active site of the enzyme and the flow of electrons toward the side chain carbonyl group of DOX. This model was strengthened by the fact that high concentrations of PTX (i.e., concentrations that would not be reached in the heart under standard clinical conditions) competed with low affinity ($^2K_m > 6\,\mu M$) for the active site of the reductases, thereby displacing DOX and inhibiting DOXOL formation [26, 27]. Of particular note is that PTX never increased ROS formation in human myocardial strips exposed to DOX [27]. This suggests that the higher than expected cardiotoxicity of DOX-PTX combinations would be caused quite specifically by the stimulation of DOXOL formation, as one would expect if DOXOL served as the long-lived anthracycline reservoir with a higher potency toward many cellular targets. Equally important is the observation that DCT stimulated DOXOL formation—but not ROS formation—by the same allosteric mechanisms and with the same efficacy as those described with PTX (Table 2). This raises concerns about the actual safety of DOX-DCT doublets, and suggests that also DCT might precipitate cardiac events if it were combined with cumulative doses of DOX higher than those adopted in the available clinical studies.

The case of DOX-taxane combinations served an excellent opportunity to probe the metabolic foundations and clinical correlates of replacing DOX with analogues characterized by a defective conversion to DOXOL. Once again, the "old" EPI proved useful to challenge "new" hypotheses. Clinical trials of EPI immediately followed by PTX had shown that CHF only occurred at cumulative doses of >900 mg of EPI/m^2; this was more than twice as high as the maximum cumulative dose of DOX that could be safely administered in combination with PTX [79].

Likewise, EPI immediately followed by DCT only caused minor and clinically manageable symptoms in a patient exposed to 870 mg of EPI/m^2 [80]. On balance, these results uncovered that combining EPI with taxanes caused essentially the same cardiotoxicity as that observed with EPI alone, as if taxanes and EPI failed to engage toxic metabolic interactions. This proved to be the case: when assessed in human myocardial samples, neither PTX nor DCT improved the kinetics of EPIOL formation (Table 2). While correlating nicely with the clinical tolerability of EPI-taxane regimens, these results confirmed that the levels of secondary alcohol metabolites determined the cardiac tolerability of anthracycline-based regimens.

5.2
Anthracyclines and Trastuzumab

p185HER2, product of the protooncogene HER2 (also known as c-erbB-2 or neu), is a transmembrane receptor tyrosine kinase belonging to the epidermal growth factor receptor family. p185HER2 (hereafter referred to as HER2 *tout court*) is overexpressed in approximately 15–25% of human breast cancers, and such alteration associates with a poor prognosis. Pivotal studies of women with HER2$^+$ metastatic breast cancer showed that a great therapeutic improvement could be obtained with Trastuzumab, a humanized monoclonal antibody targeted against the extracellular domain of HER2. Unfortunately, however, it turned out that also the heart expressed a definite level of HER2; the clinical use of Trastuzumab therefore associated with a probability of cardiac events, which increased dramatically if the patients received concomitant 300–360 mg of DOX/m^2 [81].

A retrospective analysis of the cardiac tolerability of Trastuzumab in the metastatic setting then showed that the probability of a cardiac event decreased significantly if DOX and Trastuzumab were administered sequentially [82]. This latter finding was confirmed by the HERA trial or other similar trials which adopted DOX followed by Trastuzumab for the adjuvant treatment of HER2$^+$ breast cancer [83, 84].

The mechanism(s) of Trastuzumab cardiotoxicity have not been fully elucidated, nor is it clear why Trastuzumab synergized strongly with concomitant DOX but much less with other chemotherapeutics (PTX, platinum compounds) or prior DOX. Studies with mice suggest that HER2 is a key player of cardiac development and an important factor in maintaining cardiac structure and function [85]. In vitro studies also show that the binding of neuregulin-1 (physiologically produced by endothelial cells) to HER2-HER4 heterodimers triggers a cascade of growth and survival signals that are important for cardiomyocytes to withstand stressor agents: in the case of an anthracycline-induced stress, neuregulin-1 prevents myofilament disorganization/degradation and necrotic or apoptotic death [85]. The fact that HER2 and HER4 localize primarily to the transverse tubules of ventricular myocytes lends support to the concept that HER2-derived signals are essential for a dynamic regulation of sarcomeric proteins under physiologic and especially pathologic conditions.

The last few years have witnessed attempts to highlight differences in the cardiac toxicities induced by Trastuzumab or DOX. The current thinking is that Trastuzumab alone induces a myocardial dysfunction (heralded by a decline of the LVEF) which develops dose-independently, usually lacks ultrastructural changes at endomyocardial biopsy, shows reversibility upon Trastuzumab withdrawal, does not recur upon rechallenge nor precipitates late sequelae upon exposure to sequential stressors. The characteristics of Trastuzumab cardiotoxicity are virtually opposite to those of DOX; therefore, the cardiotoxicity induced by DOX or Trastuzumab is referred to as type I or II, respectively (Table 3) [86].

In the light of such differences the toxic synergism of Trastuzumab with DOX might be recapitulated within the framework of a "two hits damage". With concomitant DOX and Trastuzumab, the toxicity of DOX is strongly

Table 3 Anthracyclines vs. Trastuzumab: Type I vs. Type II cardiotoxicity

	Type I Cardiotoxicity (anthracycline)	Type II Cardiotoxicity (Trastuzumab)
Clinical course, response to medication	May stabilize, but subclinical damage seems to persist; recurrence in months or years may be related to sequential cardiac stress	High likelihood of complete or near-to-complete recovery upon withdrawal and/or medication
Dose-dependence	Cumulative, "lifetime" dose-related	Dose-independent
Mechanism	Free radical formation (?), alcohol metabolite formation (?)	Elimination of HER2-related survival factors
Ultrastructure	Vacuoles, myofibrillar disarray and dropout, apoptosis and necrosis	With limited exceptions, no apparent ultrastructural abnormalities
Noninvasive cardiac testing	Decreased LVEF, global decrease in wall motion	As in type I
Effect of rechallenge	High probability of recurrent dysfunction that progresses toward treatment-resistant CHF	Increasing evidence for the safety of rechallenge
Effect of late sequential stress	High likelihood of sequential stress-related cardiac dysfunction	Low likelihood of sequential stress-related cardiac dysfunction

Adapted from [82, 86]
LVEF, left ventricular ejection fraction; CHF, congestive heart failure

amplified by elimination of the HER2-derived survival signals; with DOX followed by Trastuzumab, cardiotoxicity is less severe and reflects an overlap of the "mild" toxicity of Trastuzumab with the ongoing subclinical damage that is induced by a residual intramyocardial pool of anthracycline. There are clinical correlates to support such an interpretation. In the case of concomitant anthracycline and Trastuzumab, studies in the metastatic settings show that the incidence and severity of cardiac events may be diminished by replacing DOX with analogs or formulations that produce fewer toxic species, like EPI [87] and uncoated or pegylated liposomal DOX [14]. In the case of anthracycline followed by Trastuzmab, studies in the adjuvant setting show that EPI can be administered at cumulative doses higher than equiactive to DOX, as one would expect if the defective conversion of EPI to EPIOL diminished the long-lived anthracycline reservoir that mediated an ongoing subclinical damage [83].

The cardiac toxicity of Trastuzumab single agent or in combination with anthracyclines is just one example of the untoward effects caused by drugs targeted to "specific" tumor proteins. While popularly referred to as "magic bullets", many such drugs are under scrutiny for their tendency to cause undesired effects in normal tissues. In the case of the heart, or of the cardio-vascular system in general, there is an increasing concern about the actual safety of small tyrosine kinase inhibitors like lapatinib (targeted to HER2 and HER1), imatinib (targeted to the leukaemogenic oncogene Bcr-Abl but also to the receptor protein *c*-Kit of gastrointestinal sarcomas), sunitinib (targeted to the vascular endothelial growth factor/VEGF receptor-2). Similar concerns extend to anti-VEGF monoclonal antibodies such as bevacizumab. There is more than one reason to foresee many new clinical paradigms of cardiotoxi-city if these agents were considered for combination with anthracyclines. As pointed out, chemists, pharmacologists, and oncologists might better serve cancer patients by studying the terminal ballistics of new agents rather than persisting in a search for wonder "magic bullets" [88].

References

1. Minotti G, Menna P, Salvatorelli E, Cairo G, Gianni L (2004) Pharmacol Rev 56:185
2. Steinherz LJ, Steinherz PG, Tan CTC, Heller G, Murphy L (1991) J Am Med Ass 266:1672
3. Swain SM, Whaley FS, Ewer MS (2003) Cancer 97:2869
4. Stewart DJ, Grewaal D, Green RM, Mikhael N, Goel R, Montpetit VA, Redmond MD (1993) Anticancer Res 13:1945
5. Meissner K, Sperker B, Karsten C, Zu Schwabedissen HM, Seeland U, Bohm M, Bien S, Dazert P, Kunert-Keil C, Vogelgesang S, Warzok R, Siegmund W, Cascorbi I, Wendt M, Kroemer HK (2002) J Histochem Cytochem 50:1351
6. Lipshultz SE, Lipsitz SR, Sallan SE, Simbre VC 2nd, Shaikh SL, Mone SM, Gelber RD, Colan SD (2002) J Clin Oncol 20:4517
7. Singal PK, Iliskovic N (1998) N Engl J Med 339:900
8. Cusack BJ, Young SP, Driskell J, Olson RD (1993) Cancer Chemother Pharmacol 32:53
9. Legha SS, Benjamin RS, Mackay B, Legha SS, Benjamin RS, Mackay B, Ewer M, Wallace S, Valdivieso M, Rasmussen SL, Blumenschein GR, Freireich EJ (1982) Ann Intern Med 96:133
10. Lipshultz SE, Giantris AL, Lipsitz SR, Kimball Dalton V, Asselin BL, Barr RD, Clavell LA, Hurwitz CA, Moghrabi A, Samson Y, Schorin MA, Gelber RD, Sallan SE, Colan SD (2002) J Clin Oncol 20:1677
11. Batist G, Ramakrishnan G, Rao CS, Chandrasekharan A, Gutheil J, Guthrie T, Shah P, Khojasteh A, Nair MK, Hoelzer K, Tkaczuk K, Park YC, Lee LW (2001) J Clin Oncol 19:1444
12. Harris L, Batist G, Belt R, Rovira D, Navari R, Azarnia N, Welles L, Winer E (2002) Cancer 94:25
13. Batist G, Harris L, Azarnia N, Lee LW, Daza-Ramirez P (2006) Anticancer Drugs 17:587
14. Batist G (2007) Cardiovasc Toxicol 7:72
15. Cortes J, Estey E, O'Brien S, Giles F, Shen Y, Koller C, Beran M, Thomas D, Keating M, Kantarjian H (2001) Cancer 92:7
16. Fassas A, Buffels R, Anagnostopoulos A, Gacos E, Vadikolia C, Haloudis P, Kaloyan-nidis P (2002) Br J Haematol 116:308

17. O'Byrne KJ, Thomas AL, Sharma RA, DeCatris M, Shields F, Beare S, Steward WP (2002) Br J Cancer 87:15
18. Jeyaseelan R, Poizat C, Wu HY, Kedes L (1997) J Biol Chem 272:5828
19. Arai M, Tomaru K, Takizawa T, Sekiguchi K, Yokoyama T, Suzuki T, Nagai R (1998) J Mol Cell Cardiol 30:243
20. Zhou S, Heller LJ, Wallace KB (2001) Toxicol Appl Pharmacol 175:60
21. Marcillat O, Zhang Y, Davies KJ (1989) Biochem J 259:181
22. Minotti G, Recalcati S, Menna P, Salvatorelli E, Corna G, Cairo G (2004) Methods Enzymol 378:340
23. Rush JD, Koppenol WH (1990) FEBS Lett 275:114
24. Kalyanaraman B, Joseph J, Kalivendi S, Wang S, Konorev E, Kotamraju S (2002) Mol Cell Biochem 234/235:119
25. Li T, Danelisen I, Singal PK (2002) Mol Cell Biochem 232:19
26. Salvatorelli E, Menna P, Cascegna S, Liberi G, Calafiore A, Gianni L, Minotti G (2006) J Pharmacol Exp Ther 318:424
27. Salvatorelli E, Menna P, Gianni L, Minotti G (2007) J Pharmacol Exp Ther 320:790
28. Olson RD, Mushlin PS (1990) Doxorubicin cardiotoxicity: analysis of prevailing hypotheses. FASEB J 4:3076
29. Minotti G, Parlani M, Salvatorelli E, Menna P, Cipollone A, Animati F, Maggi CA, Manzini S (2001) Br J Pharmacol 134:1271
30. Charlier HA Jr, Olson RD, Thornock CM, Mercer WK, Olson DR, Broyles TS, Muhlestein DJ, Larson CL, Cusack BJ, Shadle SE (2005) Mol Pharmacol 67:1505
31. Gambliel HA, Burke BE, Cusack BJ, Walsh GM, Zhang YL, Mushlin PS, Olson RD (2002) Biochem Biophys Res Commun 291:433
32. Minotti G, Recalcati S, Mordente A, Liberi G, Calafiore AM, Mancuso C, Preziosi P, Cairo G (1998) FASEB J 12:541
33. Cairo G, Recalcati S, Pietrangelo A, Minotti G (2002) Free Radic Biol Med 32:1237
34. Licata S, Saponiero A, Mordente A, Minotti G (2000) Chem Res Toxicol 13:414
35. Minotti G, Saponiero A, Licata S, Menna A, Calafiore AM, Teodori G, Gianni L (2001) Clin Cancer Res 7:1511
36. Forrest GL, Gonzalez B, Tseng W, Li X, Mann J (2000) Cancer Res 60:5158
37. Olson LE, Bedja D, Alvey SJ, Cardounel AJ, Gabrielson KL, Reeves RH (2003) Cancer Res 63:660
38. Egawa T, Shimada H, and Ishimura Y (2000) Formation of compound I in the reaction of native myoglobins with hydrogen peroxide. J Biol Chem 275:34858
39. Menna P, Salvatorelli E, Giampietro R, Liberi G, Teodori G, Calafiore AM, Minotti G (2002) Chem Res Toxicol 15:1179
40. Reszka KJ, McCormick ML, Britigan BE (2001) Biochemistry 40:15349
41. Cartoni A, Menna P, Salvatorelli E, Braghiroli D, Giampietro R, Animati F, Urbani A, Del Boccio P, Minotti G (2004) J Biol Chem 13:5088
42. Reszka KJ, Wagner BA, Teesch LM, Britigan BE, Spitz DR, Burns CP (2005) Inact Cancer Res 65:6346
43. Menna P, Salvatorelli E, Minotti G (2007) J Pharmacol Exp Ther 322:408
44. Ladas EJ, Jacobson JS, Kennedy DD, Teel K, Fleischauer A, Kelly KM (2004) J Clin Oncol 22:517
45. Dulin B, Abraham WT (2004) Am J Cardiol 93:3B
46. Kalay N, Basar E, Ozdogru I, Er O, Cetinkaya Y, Dogan A, Inanc T, Oguzhan A, Eryol NK, Topsakal R, Ergin A (2006) J Am Coll Cardiol 48:2258
47. Florenzano F, Salman P (2007) J Am Coll Cardiol 49:2142
48. Hasinoff BB, Patel D, Wu X (2007) Cardiovasc Toxicol 7:19

49. Scully R, Lipshultz SE (2007) Cardiovasc Toxicol 7:122
50. Swain SM, Whaley FS, Gerber MC, Weisberg S, York M, Spicer D, Jones SE, Wadler S, Desai A, Vogel C, Speyer J, Mittelman A, Reddy S, Pendergrass K, Velez-Garcia E, Ewer MS, Bianchine JR, Gams RA (1997) J Clin Oncol 15:1318
51. Swain SM, Vici P (2004) J Cancer Res Clin Oncol 130:1
52. Kim E, Giri SN, Pessah IN (1995) Toxicol Appl Pharmacol 130:57
53. Sterba M, Popelová O, Simunek T, Mazurová Y, Potácová A, Adamcová M, Kaiserová H, Ponka P, Gersl V (2006) J Pharmacol Exp Ther 319:1336
54. Kaiserová H, Simunek T, Sterba M, den Hartog GJ, Schröterová L, Popelová O, Gersl V, Kvasnicková E, Bast A (2007) Cardiovasc Toxicol 7:145
55. Anderlini P, Benjamin RS, Wong FC, Kantarjian HM, Andreeff M, Kornblau SM, O'Brien S, Mackay B, Ewer MS, Pierce SA (1995) J Clin Oncol 13:2827
56. Crivellari D, Lombardi D, Corona G, Massacesi C, Talamini R, Sorio R, Magri MD, Lestuzzi C, Lucenti A, Veronesi A, Toffoli G (2006) Ann Oncol 17:807
57. Toffoli G, Sorio R, Basso B, Aita P, Corona G, Ruolo G, Boiocchi M (2004) J Chemother 16:193
58. Innocenti F, Iyer L, Ramirez J, Green MD, Ratain MJ (2001) Drug Metab Dispos 29:686
59. Bonadonna G, Gianni L, Santoro A, Bonfante V, Bidoli P, Casali P, Demicheli R, Valagussa P (1993) Ann Oncol 4:359
60. Salvatorelli E, Guarnieri S, Menna P, Liberi G, Calafiore AM, Mariggio MA, Mordente A, Gianni L, Minotti G (2006) J Biol Chem 281:10990
61. Arcamone F (1985) Cancer Res 45:5995
62. Altan N, Chen Y, Schindler M, Simon SM (1998) J Exp Med 187:1583
63. DeFeo-Jones D, Garsky VM, Wong BK, Feng DM, Bolyar T, Haske ll K, Kiefer DM, Leander K, McAvoy E, Lumma P, Wai J, Senderak ET, Motzel SL, Keenan K, Van Zwieten M, Lin JH, Freidinger R, Huff J, Oliff A, Jones RE (2000) Nat Med 6:1248
64. Seymour LW, Ferry DR, Anderson D, Hesslewood S, Julyan PJ, Poyner R, Doran J, Young AM, Burtles S, Kerr DJ (2002) J Clin Oncol 20:1668
65. Olson RD, Headley MB, Hodzic A, Walsh GM, Wingett DG (2007) Int Immunopharmacol 7:734
66. Arcamone F, Animati F, Berettoni M, Bigioni M, Capranico G, Casazza AM, Caserini C, Cipollone A, De Cesare M, Franciotti M, Lombardi P, Madami A, Manzini S, Monteagudo E, Polizzi D, Pratesi G, Righetti SC, Salvatore C, Supino R, Zunino F (1997) J Natl Cancer Inst 89:1217
67. Pratesi G, De Cesare M, Caserini C, Perego P, Dal Bo L, Polizzi D, Supino R, Bigioni M, Manzini S, Iafrate E, Salvatore C, Casazza A, Arcamone F, Zunino F (1998) Clin Cancer Res 4:2833
68. Minotti G, Licata S, Saponiero A, Menna P, Calafiore AM, Di Giammarco G, Liberi G, Animati F, Cipollone A, Manzini S, Maggi CA (2000) Chem Res Toxicol 13:1336
69. Minotti G, Parlani M, Salvatorelli E, Menna P, Cipollone A, Animati F, Maggi CA, Manzini S (2001) Br J Pharmacol 134:1271
70. Sacco G, Giampietro R, Salvatorelli E, Menna P, Bertani N, Graiani G, Animati F, Goso C, Maggi CA, Manzini S, Minotti G (2003) Br J Pharmacol 139:641
71. Schrijvers D, Bos AM, Dyck J, de Vries EG, Wanders J, Roelvink M, Fumoleau P, Bortini S, Vermorken JB (2002) Ann Oncol 13:385
72. Caponigro F, Willemse P, Sorio R, Floquet A, van Belle S, Demol J, Tamburo R, Comandino A, Capriati A, Adank S, Wanders J (2005) New Drugs 23:85
73. Fiedler W, Tchen N, Bloch J, Fargeot P, Sorio R, Vermorken JB, Collette L, Lacombe D, Twelves C (2006) Eur J Cancer 42:200

74. Reszka KJ, Britigan LH, Britigan BE (2005) J Pharmacol Exp Ther 315:283
75. Schimmel JMK, Richelb DJ, van den Brink RBA, Guchelaard H-J (2004) Cancer Treat Rev 30:181
76. Gianni L, Munzone E, Capri G, Fulfaro F, Tarenzi E, Villani F, Spreafico C, Laffranchi A, Caraceni A, Martini C, Stefanelli M, Valagussa P, Bonadonna G (1995) J Clin Oncol 13:2688
77. Perotti S, Cresta G, Grasselli G, Capri G, Minotti G, Gianni L (2003) Expert Opin Drug Saf 2:59–71
78. Nabholtz JM, Falkson C, Campos D, Szanto J, Martin M, Chan S, Pienkowski T, Zaluski J, Pinter T, Krzakowski M, Vorobiof D, Leonard R, Kennedy I, Azli N, Murawsky M, Riva A, Pouillart P (2003) J Clin Oncol 21:968
79. Gennari A, Salvadori B, Donati S, Bengala C, Orlandini C, Danesi R, Del Tacca M, Bruzzi P, Conte PF (1999) J Clin Oncol 17:3596
80. Pagani O, Sessa C, Nole F, Crivellari D, Lombardi D, Thurlimann B, Hess D, Borner M, Bauer J, Martinelli G, Graffeo R, Zucchetti M, D'Incalci M, Goldhirsch A (2000) Ann Oncol 11:985
81. Slamon DJ, Leyland-Jones B, Shak S, Fuchs H, Paton V, Bajamonde A, Fleming T, Eiermann W, Wolter J, Pegram M, Baselga J, Norton L (2001) N Engl J Med 344:783
82. Guarneri V, Lenihan DJ, Valero V, Durand JB, Broglio K, Hess KR, Michaud LB, Gonzalez-Angulo AM, Hortobagyi GN, Esteva FJ (2006) J Clin Oncol 24:4107
83. Suter TM, Procter M, van Veldhuisen DJ, Muscholl M, Bergh J, Carlomagno C, Perren T, Passalacqua R, Bighin C, Klijn JGM, Ageev FT, Hitre E, Groetz J, Iwata H, Knap M, Gnant M, Muehlbauer S, Spence A, Gelber RD, Piccart-Gebhart MJ (2007) J Clin Oncol 25:3859
84. Gianni L, Salvatorelli E, Minotti G (2007) Cardiovasc Toxicol 7:67
85. Peng X, Chen B, Lim CC, Sawyer DB (2005) Mol Interv 5:163
86. Ewer MS, Lippman SM (2005) J Clin Oncol 23:2900
87. Untch M, Eidtmann H, du Bois A, Meerpohl HG, Thomssen C, Ebert A, Harbeck N, Jackisch C, Heilman V, Emons G, Wallwiener D, Wiese W, Blohmer JU, Höffken K, Kuhn W, Reichardt P, Muscholl M, Pauschinger M, Langer B, Lück HJ (2004) Eur J Cancer 40:988
88. Maitland ML, Ratain MJ (2006) Ann Intern Med 145:702

Top Curr Chem (2008) 283: 45–71
DOI 10.1007/128_2007_2
© Springer-Verlag Berlin Heidelberg
Published online: 21 November 2007

Daunomycin-TFO Conjugates for Downregulation of Gene Expression

Massimo L. Capobianco[1] (✉) · Carlo V. Catapano[2]

[1]ISOF-CNR, Via Gobetti 101, 40129 Bologna, Italy
capobianco@isof.cnr.it

[2]Laboratory of Experimental Oncology, IOSI, Via Vela 6, 6500 Bellinzona, Switzerland

Abstract Daunomycin has shown interesting properties as a stabilizing agent for the anti-gene methodology. This approach consists of targeting a polypurine region of a given gene, with a triplex forming oligonucleotide (TFO), realizing a triple helix complex (triplex), with the aim of down-regulating gene expression. This chapter describes the basic principles of the triplex approach, the chemistry underlining the binding of daunomycin to oligonucleotides, and some results of gene-inhibition obtained with daunomycin-TFO conjugates with different targets.

Keywords Anti-gene · c-myc · Daunomycin · HIV · Oligonucleotides · Triplex DNA

Abbreviations

DOTAP	*N*-(2,3-dioleoyloxy-1-propyl)trimethylammonium
DTT	1,4-dithio-DL-threitol
T_m	Melting temperature
TEAA	Triethylammonium acetate
TFO	Triplex forming oligonucleotide
Tlc	Thin-layer chromatography
TMSOTf	Trimethylsilyl trifluoromethanesulphonate

1
The Anti-Gene Methodology

The existence of a 1 : 2 complex between poly-A and poly-U was first described by Fenselfeld and co-workers [1] in 1957, shortly after the discovery of the double helix structure [2]. This discovery remained largely unexploited until the end of the 1980s when pioneers like P. B. Dervan [3] and C. Hélène [4] began their research in the field of artificial endonucleases. Their work opened the way to research aimed at downregulating the expression of a given gene by interfering with the transcription process trough the formation of a triple-helix (triplex) directed by a triplex-forming oligonucleotide (TFO), thereby the topic began to be indicated as anti-gene methodology [5]. This methodology relies on the possibility to specifically bind an oligonucleotide in the major groove of the targeted gene through specific hydrogen bonds formed between the bases of the TFO and a homo-purine region (Hoogsteen's bonds) present in one of the strands of the gene. The patterns of the most common triplets formed by the bases of the TFO and those of the targeted duplex are depicted in Fig. 1.

When the sequence of the gene of interest is known, it is relatively easy to find a homo-purine tract suitably long (10–30 bases) to which address a series of TFOs specific for that region. Fortunately, homo-purine tracts seem to be located inside important regulatory regions of the genes [6] with high frequency.

To be selective toward the intended target, the TFO should be 17 bases long [7], but this holds only on a statistical base. Nowadays it is possible to check the existence of potential molecular targets in unwanted genes by a simple search in suitable databases. Also, if we consider that at any given moment only a subset of the entire genome is translated, we can presume that unexpressed genes will be less prone to bind the TFO than the target, and this probably reduces the length required for selectivity.

In principle, the anti-gene methodology is quite straightforward: the idea is to inhibit the synthesis of a given protein interfering with its transcription by binding a TFO to a suitable homo-purine tract in the selected gene. For real applications, however, many obstacles must be faced, including uptake of

Fig. 1 Triplets in triplex. Scheme of some Hoogsteen's (*left side*) and reverse Hoogsteen's ▶ (*right side*) hydrogen bonds between the bases of the TFO (*top of each triplet*) and the Watson-Crick base pairs. The *arrows* indicate the orientation of the strands. When the TFO binds parallel to the polypurine sequence (Hoogsteen's bonds) the adenine can be bound by a thymine or by an inosine, while the guanine can be targeted by another guanine or by a protonated cytidine. In the antiparallel series (reverse Hoogsteen's bonds), adenine can be recognized either by another adenine or by a thymine, while guanine can be bound by another guanine or by a protonated cytidine. In the latter case, the strength of the bond is pH-dependent

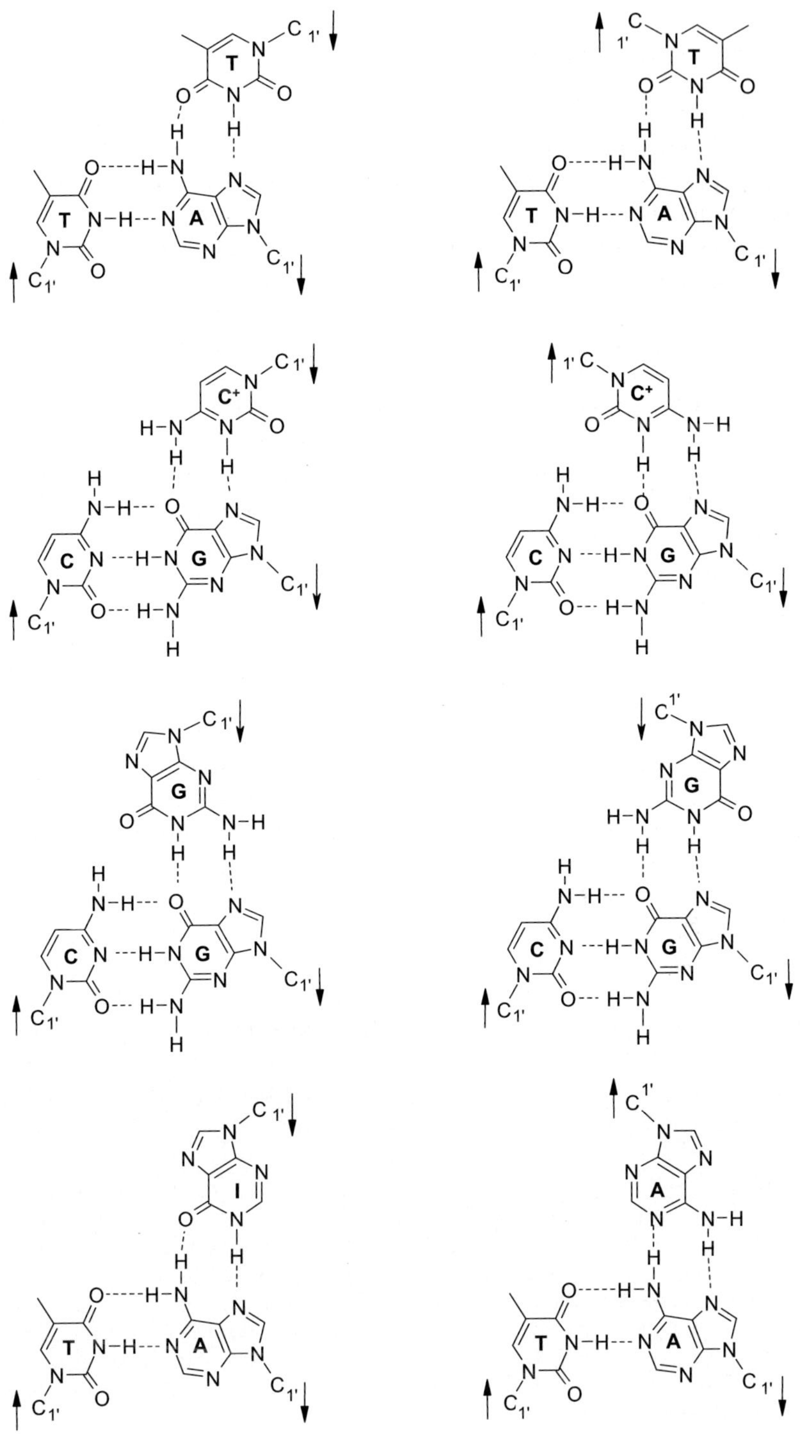

TFOs in cells, susceptibility of oligonucleotides toward nucleases, availability of the gene to binding by TFOs [8, 9], competition between the stability of the triplex and strength of the transcription machinery, effects of the positioning of the triplex along the gene on the inhibition of transcription, life time of the targeted protein, importance of the inhibition of the targeted protein to the aim of the therapy, and behavior of the system in vivo. A full treatment of such issues is out of the scope of this chapter but many papers dealing with the anti-gene strategy report a certain degree of success despite the obstacles mentioned above.

2
Synthesis of Daunomycin-Oligonucleotide Conjugates

Whatever the mechanism of interference with the transcription machinery may be, it can be predicted that a more stable triplex will give a higher inhibition of the synthesis of the encoded RNA and protein. Since the first experiments, a common way to stabilize a triplex was the linkage of an intercalating molecule to the TFO.

Our own approach was that of exploiting the intercalating properties of the anthraquinone moiety to stabilize a triplex. A search of the literature showed that only one attempt had been made before, with little improvement of the binding capability of the TFO [10]. In that study, daunomycin was conjugated with the TFO by linking it to the amino sugar. However, a similar arrangement forces the amino sugar and ring A (Fig. 2) to be located in the major groove and this is not compatible with the known structure of daunomycin-DNA complexes [11] where anthracyclines intercalate into DNA with the long axis of the aglycone moiety nearly perpendicular to the long axis of the adjacent base pairs, and with the D ring protruding into the major groove while the amino sugar fits into the minor groove and further stabilizes the complex.

On the other hand, the preferred orientation of daunomycin in complexes with DNA would be respected upon linking the TFO to the D ring. This can

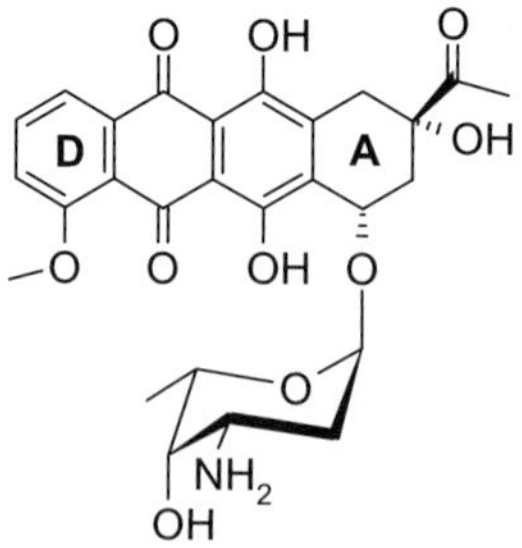

Fig. 2 Daunomycin

be achieved replacing the methyl group with a flexible spacer having a good leaving group at the other end, and condensing it with a thiophosphate group on the TFO (Scheme 1) [10].

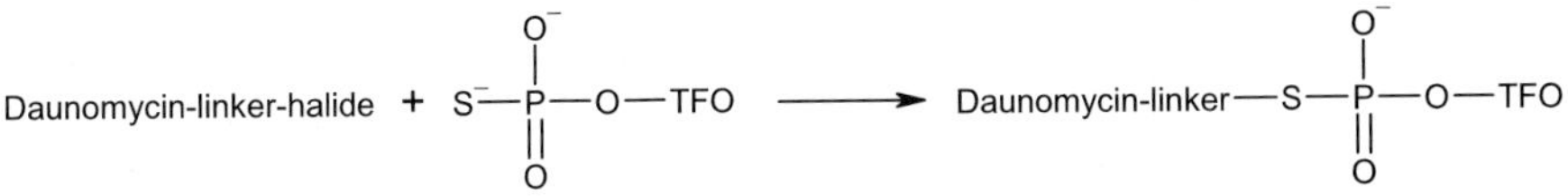

$$\text{Daunomycin-linker-halide} \; + \; \text{S}\overset{\overset{\displaystyle O^-}{|}}{\underset{\underset{\displaystyle O}{||}}{\text{P}}}\text{O—TFO} \quad \longrightarrow \quad \text{Daunomycin-linker—S}\overset{\overset{\displaystyle O^-}{|}}{\underset{\underset{\displaystyle O}{||}}{\text{P}}}\text{O—TFO}$$

Scheme 1 Post-synthetic conjugation of oligonucleotide (TFO) with daunomycin derivatives

Our first approach to prepare the ω-iodohexyl derivative of daunomycin is summarized in Scheme 2 [12].

Daunomycin hydrochloride 1 was allowed to react with allyloxycarbonyl chloride in dichloromethane in the presence of pyridine and the resulting compound 2 was hydrolyzed with acid. Demethylation of 3 gave carminomycinone 5 [13]. Alkylation of carminomycinone was performed at a high dilution, partly for the low solubility of carminomycinone, partly to avoid the bis-substitution, using six to eight equivalents of di-iodohexane and two equivalents of fresh Ag$_2$O for three or more days adding 25% of fresh Ag$_2$O every day. Two main compounds were formed and separated by silica gel chromatography. Reaction of the amino sugar moiety with p-nitrobenzoyl chloride gave two anomers that were isolated by crystallization being 8 (the more abundant and the first to crystallize) the anomer that reacts faster in the subsequent reaction with alkylated carminomycinone. The reaction took less than 1 h to give 9, that was purified by chromatography.

Deprotection of 9 by a two-step procedure produced 10, isolated by lyophilization. This compound proved to be stable in that form for several years in the freezer.

Working with these compounds, we noticed that one of the by-products of the coupling reaction (from 2 to 5%) was probably derived from N-alkylation of the conjugate with the halide of a second daunomycin reagent. Also 10 was found to be unstable in the presence of traces of acid. These problems were solved by protecting the amino group of daunomycin as a trifluoroacetyl derivative and removing the protecting group after the conjugation with the TFO as shown in Scheme 3 [14].

N-trifluoroacetyl-daunosamine was allowed to react with p-nitrobenzoyl-chloride to give 86% of 1,4-bis-p-nitrobenzoyl-N-trifluoroacetyl daunosamine with an α/β ratio of 9/1, the α-isomer (11) being obtained after crystallization of the mixture from chloroform. The glycosylation of 6 to give 13 was performed upon treatment of 11 with TMSOTf in dichloromethane/ether at $-40\,^\circ$C to form the carbocation that was then reacted with a solution of 6 in dichloromethane (variable yields between 50 and 70% after chromatographic

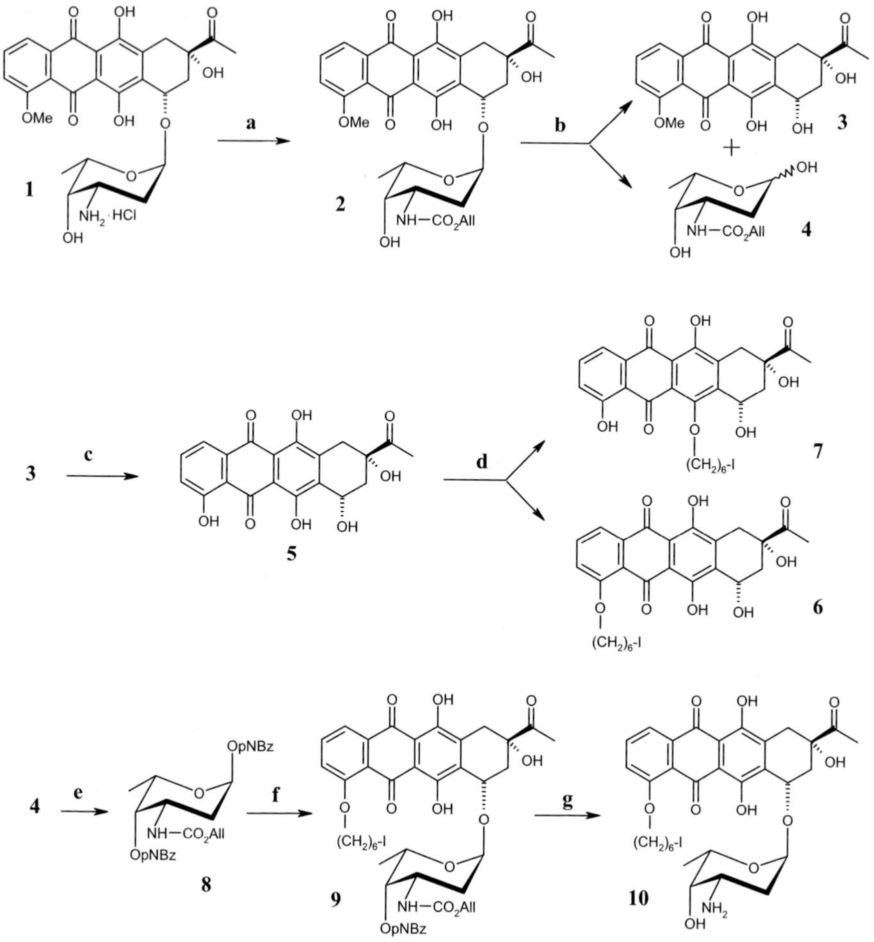

Scheme 2 Early synthesis of 4-O-(6-Iodohexyl)-daunomycin. **a** Allyloxycarbonylchloride in CH_2Cl_2/Py (yield 97.8%). **b** dioxane, aq. HCl (yield 98% of **3**, 99% of **4**). **c** $AlCl_3$ in refluxing CH_2Cl_2 (yield 70%). **d** 1,6-diiodohexane, Ag_2O, in refluxing $CHCl_3$, 3–5 days (yield 15% of **6**, 13% of **7**). **e** p-nitrobenzoylchloride in py (yield 41% α/β = 2.3). **f 6**, TM-SOTf in CH_2Cl_2/Et_2O 4/1, 4 Å MS (yield 60%). **g** 0.5 M aq. K_2CO_3 in MeOH, (yield 58% of de-pNBz compound) then 2-methylbutiric acid, PPh_3, $Pd(PPh_3)_4$ in dark CH_2Cl_2 (yield 80% of **10**)

separation). Compound **14** was obtained by treatment of **13** with base. The glycosylation reaction was also performed starting from **3** using the same procedure. Protected carminomycin **12** was alkylated (reaction **d**) directly with a higher yield of **13** (about 40% on tlc) although problems were encountered during the purification, which reduced the yield to only 20% of pure **13** after repeated column chromatography. This latter procedure however is of interest in view of the potential availability of biosynthetic carminomycin **14**

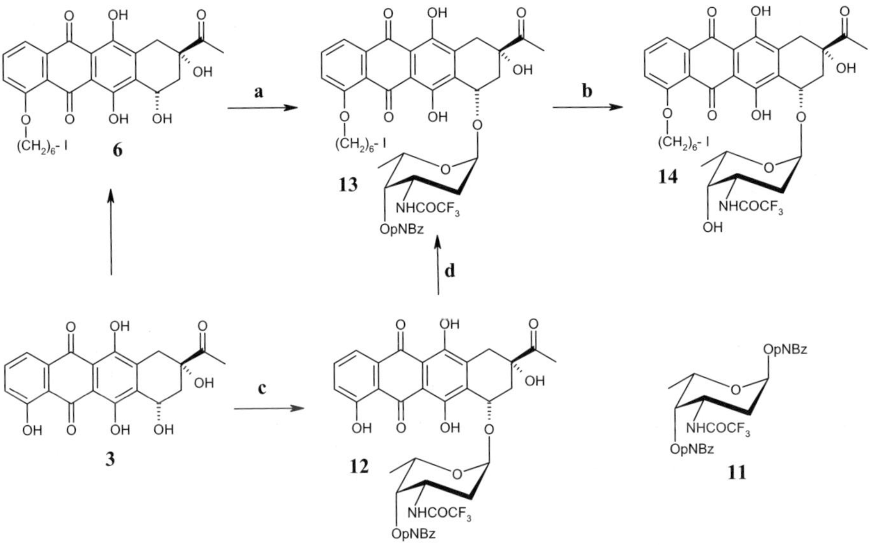

Scheme 3 Improved synthesis of 4-O-(6-Iodohexyl)-daunomycin. **a** **11**, TMSOTf in CH$_2$Cl$_2$/Et$_2$O 4/1, 4 Å MS at $-15\,°$C, then **6** from -15 to $-5\,°$C (yield 60%)

as the starting material[1]. In this case (Scheme 4), direct alkylation of trifluoroacetyl carminomycin **15** allowed a satisfactory conversion to **13** that was easily purified, with a 25–30% overall yield from **14**.

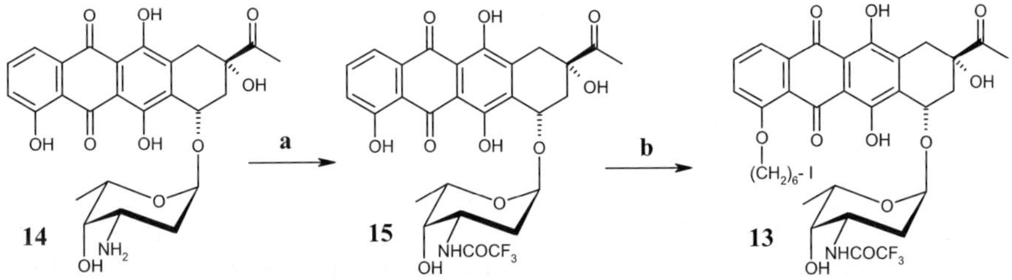

Scheme 4 Direct synthesis of 4-O-(6-Iodohexyl)-daunomycin from carminomycin

The coupling of the suitable iodoalkyl-daunomycin derivative and the 5′ (or 3′) thiophosphate-oligonucleotide was performed as shown in Scheme 5.

The 5′-thiophosphate-oligonucleotides (**15**) can be prepared using an appropriate phosphoramidite as the last coupling step in an automated oligonucleotide synthesizer, then replacing the oxidation step with a thiooxidation one. We usually employed bis-cyanoethyl-N,N-diisopropyl phos-

---

[1] From 2001 to 2003 we worked together with Biofin Laboratories, Porto Mantovano, Italy, to develop the chemistry of daunomycin-TFO conjugates. They now have carminomycin **14** in their catalog.

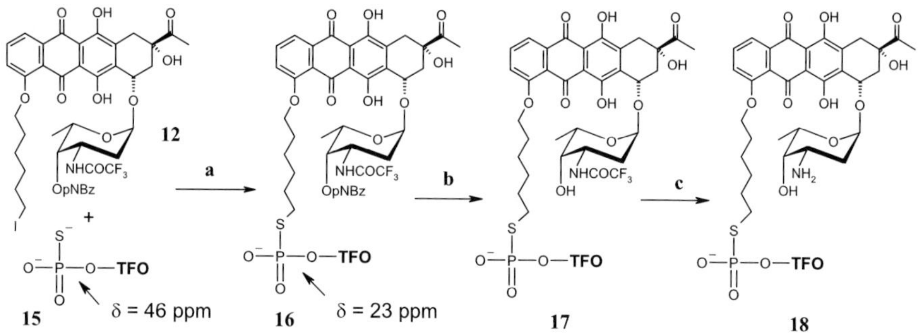

Scheme 5 Synthesis of daunomycin-TFO conjugates

phoramidite [16] as a reactant, and Beaucage's reagent [17] for the thio-oxidation, but others amidites [18] and thio-oxidizers [19] can be used as well. 3'-Thiophosphate oligonucleotides can be conjugated in the same way, even if their synthesis requires the modification of the solid support [20, 21]. (In fact, the synthesis of 5',3'-bis-daunomycin oligonucleotides starting from oligonucleotides containing a thiophosphate at each end has also been achieved.) Final deprotection of the 5' (or 3')-thiophosphate oligonucleotide was performed with a concentrated solution of hot ammonia followed by lyophilization. The compound was then converted to the sodium form by the use of an ion-exchange resin treated overnight with DTT to cleave possible thiophosphate dimers recovered by precipitation with n-BuOH and lyophilized.

General Procedure

The sodium salt of a crude 5'-PS-oligonucleotide (10 OD about 75 pure) was dissolved in 125 μL of DMF, 20 μL of water, 11 μL of 15-crown-5, 1 mg of 12 (13 can also be used) and 0.2 mg of DTT. The reaction mixture was kept in a sealed vial for 16 h in a thermostat at 45 °C. The success of the conjugation can be easily monitored by reversed-phase HPLC. The crude mixture was separated from excess of iododerivatives by diluting it with 150 μL of water and extracting the unreacted iododerivative with CH_2Cl_2. The aqueous phase was cooled to 0 °C, then 1 M aqueous NaOH was added to a final concentration of 0.06 M (the conjugate turns violet).

Under these conditions, the p-nitrobenzoyl group is easily cleaved (5–10 min), followed by the removal of the trifluoroacetyl protecting group (in 70–90 min) to give compound 18, whose retention time is sharply different from that of the protected precursors. The reaction is quenched by neutralizing the pH with acetic acid (the solution turns again red-orange), then the pure conjugate is recovered after purification on a reversed-phase C-18 silica-gel column using a gradient of acetonitrile (0–30%) in a 0.1 M of triethylammonium acetate (TEAA) aqueous buffer. Usually, 5–6 OD of pure conjugate was obtained [14].

The presence of the correct thiophosphate group in the oligonucleotide and its correct ratio with the other phosphate linkages can be easily checked by performing a ^{31}P-NMR analysis (0.5 μmol are needed). The chemical shift of the terminal thiophosphate changes from 46 ppm in the initial monoester to 23 ppm in the conjugated compound (thiophosphate diester).

3
Physico-Chemical Properties of Daunomycin-Oligonucleotide Conjugates

The higher lipophilicity of the conjugates, as compared to the parent oligonucleotides, was exploited for the purification on reversed-phase (C-18) chromatography. In the purification of **dauno-C2** (see Table 2), eluting with a 0.1 M solution of TEAA and a linear gradient from 5 to 30% of acetonitrile in 30 min, the conjugate could easily be separated by HPLC from unreacted oligonucleotides (and failed sequences) whose retention times were significantly shorter than that of the conjugate (retention time of **16** was 27 min,

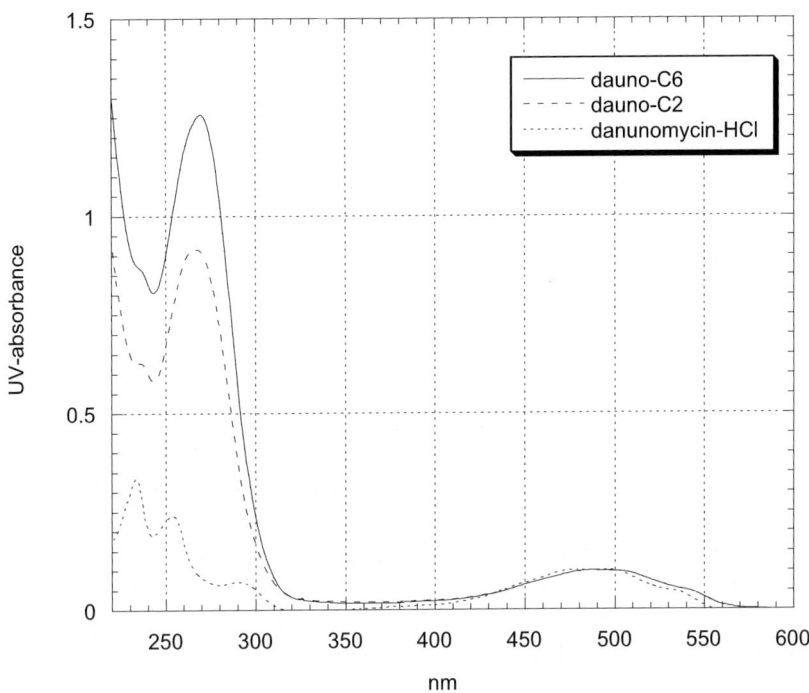

Fig. 3 UV profile of daunomycin-HCl and two dauno-oligonucleotides. The concentration of both compounds was 8 μM in water. (Conjugates are described in Table 2)

that of **17** was 23 min while that of the full deprotected conjugate **18** was 17 min). Compounds like **13**, **14**, and their hydrolyzed derivatives require a high percentage of acetonitrile (50–80%) in order to be eluted from the column.

Daunomycin conjugates are red, showing a broad absorbance in the visible region centered around 490 nm. In aqueous solutions at neutral pH, the molar absorption of the daunomycin or that of their conjugates at 490 nm is about $12\,100\ M^{-1}\ cm^{-1}$ while the contribution of the daunomycin moiety to the UV absorption of the conjugates at 260 nm can be inferred from that of the free daunomycin, and can be considered to be $20\,500\ M^{-1}\ cm^{-1}$. This value can be added to the molar absorbance of the oligonucleotide moiety that, in turn, can be calculated with the method of "the nearest neighbors" developed by Borer [22]. Thus the ratio of the absorbances at 260 and 490 is a good indicator to judge the number of incorporated daunomycin moieties into the conjugate and the degree of purity.

The linkage of daunomycin to a TFO enhances by several degrees the melting temperature (T_m) of the TFO with its duplex, as already demonstrated [12]. As shown in Table 1, the intercalating anthraquinone systems are able to stabilize the triplex (compounds **20** and **21** vs. **19**), but the amino sugar contributes increasing stability (**21** and **22**). As expected, conjugation with daunomycin increases the stability of the triplex only if the TFO is linked to the D ring (**20** and **23**).

As expected, the dependence of the stability of the triplex from pH related with the required presence of protonated cytidines for the formation of Hoogsteen's hydrogen bonds was observed (at pH 5.5 T_m were respectively: 23, 43, 35, 45, and 25 °C for the compounds from **19** to **23**).

Fluorescence of anthracyclines is quenched upon intercalation [23, 24]. Accordingly, the fluorescence of a solution containing the triplex formed by **22** and its duplex (Table 1) at pH 5.5 was almost zero [12], while a strong fluorescent signal was obtained after dissociation of the triplex upon changing the pH from 5.5 to 8.2[2]. This was proved true not only in the T,C series but also with G,T containing TFOs[3], as shown in Fig. 4 for **dauno-GT11A** in agreement with the intercalation of the anthraquinone system of the daunomycin moiety in the complexes formed in both series.

On the other hand, the fluorescence of the conjugate **dauno-CO11**, which is unable to form a triplex, is not modified after the addition of the corresponding duplex and remains high (about 35 a.u. at 590 nm). This behavior indicates that when a triplex is not formed, the intercalation of the daunomycin tagged to the TFO cannot occur. In other words, intercalation of tagged daunomycin takes place only with the formation of the triplex.

---

[2] Isolated daunomycin and daunomycin-oligonucleotide conjugate do not show any quenching of fluorescence after acidification of the solution from pH 8.2 to 5.5.

[3] See following paragraph for details on the sequences.

Table 1 Melting temperature for the triplex–duplex process

$3'$-TCC TCG **TTT CTT CTT CTT** GAA A $5'$
$5'$-AGG AGC **AAA GAA GAA GAA** CTT T-$3'$ duplex

$5'$-TTT CTT CTT CTT-$3'$ TFO

Compound		T_m in °C at pH 6.8 (ΔT_m respect to **19**)
19	$5'$TTTCTTCTTCTT$3'$ (**TFO**)	13
20		31 (+ 18)
21		27 (+ 14)
22		36 (+ 23)
23		16 (+ 3)

SP denotes a thiophosphate (–SP(O)$_2$O–) bond

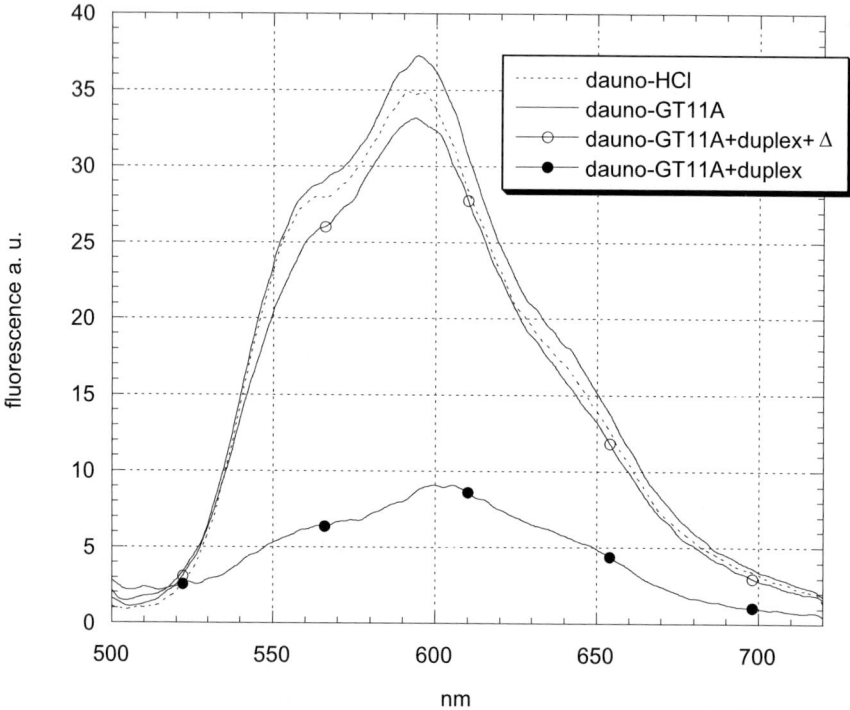

Fig. 4 Fluorescence of free daunomycin and dauno-oligonucleotides. Fluorescence measurements were taken of a 0.1 μM solution of daunomycin-HCl in 100 mM Tris-HCl pH 7.0, 50 mM NaCl, 10 mM MgCl$_2$ buffer (· · ·), **dauno-GT11A** conjugate alone at the same concentration and buffer (—), the conjugate after addition of 1 eq of corresponding duplex (-•-), and after thermal denaturation of the triplex (-o-). See Fig. 5 for the composition of conjugates

4
Biochemical and Biological Properties of Daunomycin-oligonucleotide Conjugates

While there are many examples of gene downregulation through the use of TFOs—at least in cell culture experiments—and it is clear that the binding of a daunomycin on a TFO greatly increases its ability to form a stable triplex on a given homo-purine region, until now, only a few attempts to down-regulate gene expression by use of daunomycin-oligonucleotide TFOs have been published. The reasons for this are in the difficulties encountered to obtain a clear reduction of a given RNA or protein and to attribute the results to an indisputable triple helix mechanism.

The possibility that daunomycin could act with different mechanisms, besides triplex stabilization, such as interference with the action of topoisomerases and DNA cleavage through production of free radicals have not yet been demonstrated. Unpublished experiments have shown that, in vitro, cleavages by topoisomerase II of a duplex DNA, containing a triplex binding region, occurs at different positions in the absence or in the presence of an appropriate daunomycin-TFO conjugate, but it is still unclear if this can lead to biological consequences. Also, preliminary experiments aimed at finding strand breakage of the duplex targeted region by free radicals, generated from the anthraquinone system after the triplex formation, have so far demonstrated to be inconclusive.

4.1
PPT of HIV-1 as Target

The genome of HIV-1 contains a well-conserved poly-purine/poly-pyrimidine tract, called PPT, present both in *nef* and *pol* genes [25], that could be a valuable target to fight HIV. It has been shown that the PPT sequence is accessible in the chromatin structure to an oligonucleotide to form a triple-helical structure [8]. Finding a suitable TFO able to strongly bind this region could eventually lead to a reduced DNA transcription and translation, as well as to irreparable DNA cleavage (by enzymes) followed by cellular death. This approach could then represent one of the few alternatives to recognize and destroy latently infected cells.

We have studied the hybridization properties of a series of different conjugates on duplexes of different length containing the PPT region [14, 26], see Table 2.

To assess the stability of the triplexes formed with **duplex-3**, conjugates of C2, C6 and G6 with non glycosylated 6 (derivatives A) and 7 (derivatives B) were synthesized. The purification of G6-A and G6-B was hampered by the formation of very stable aggregates (probably derived from self-association of the G_6 regions somewhat stabilized by the presence of the anthraquinone moiety) thus these series were not pursued further. Conjugates of the other two series (**C2-A, C2-B, C6-A, C6-B**) showed the formation of retarded species on gel electrophoresis ran at 15 °C at a concentration of 10 μM and pH 7.0 whereas in similar conditions unconjugated oligo C2, C6 and G6 did not show triplex formation even at 4 °C [27][4].

Using ^{32}P-labelled **duplex-1** and **duplex-2** at pH 7.2, we found a T_m of 30 and 34 °C for **dauno-C2** and **dauno-C6**, respectively [14], corresponding to an EC_{50} at 20 °C of 2.6 and 1.6 μM, respectively. This means a more than three times increased stability for **dauno-C6** compared to that of unconjugated C6 and even larger stabilization for **dauno-C2**, considering that C2 alone did not show any triplex formation even at a concentration of 40 μM! Notwithstand-

---

[4] This same sequence had been reported to form a triple helix with T_m of 34 °C at pH 6.2 and at lower concentration.

Table 2 List of oligonucleotides and compounds used on the **PPT region** of HIV-1

Duplex-1   (49-mer)

$^{5'}$CAAGGCAGCTGTAGATCTTAGCCACTTTTT**AAAAGAAAAGGGGGGA**CTG$^{3'}$

$^{3'}$GTTCCGTCGACATCTAGAATCGGTGAAAAA**TTTTCTTTTCCCCCCT**GAC$^{5'}$

Duplex-2   (29-mer)

$^{5'}$CCACTTTTT**AAAAGAAAAGGGGGGA**CTGG$^{3'}$

$^{3'}$GGTGAAAAA**TTTTCTTTTCCCCCCT**GACC$^{5'}$

Duplex-3   (25-mer)

$^{5'}$TTTTT**AAAAGAAAAGGGGGGA**CTGG$^{3'}$

$^{3'}$AAAAA**TTTTCTTTTCCCCCCT**GACC$^{5'}$

TFOs

$^{5'}$TTTT$\underline{C}$TTTT$\underline{CC}^{3'}$ (where $\underline{C}$ = ^{5Me}Cy)   **C2**

$^{5'}$TTTT$\underline{C}$TTTT$\underline{CCCCCC}^{3'}$ (where $\underline{C}$ = ^{5Me}Cy)   **C6**

$^{5'}$TTTT$\underline{C}$TTTTGGGGGG$^{3'}$ (where $\underline{C}$ = ^{5Me}Cy)   **G6**

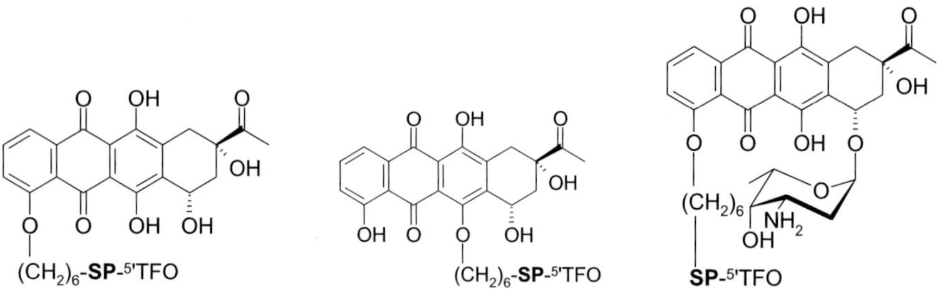

A-type conjugates   B-type conjugates   dauno conjugates

ing these improvements, the formed triplexes were deemed to be not stable
enough at physiological conditions.

4.2
The c-myc Promoter as a Target

The positive results reported so far encouraged us to take advantage of the unique DNA-binding and triplex-stabilizing properties of anthracyclines and synthesize conjugates that could bind with high affinity to a critical sequence in the promoter of a prominent human oncogene, the *c-myc* gene. The *c-myc* gene is amplified, translocated, and over-expressed in many cancers [28–30]. In the past, we and others investigated various means, including the triplex DNA-based approach, to down-regulate the expression of this oncogene in cancer cells. TFOs directed to regulatory sequences in the *c-myc* gene were shown to inhibit binding of transcription factors and transcription in vitro, promoter reporter activity and gene expression in cells [31–35]. We found that GT-rich TFOs directed to a sequence near the P2 promoter were particularly effective in inhibiting *c-myc* expression in cancer cells [36–38]. Therefore, in the attempt to increase the activity of TFOs by enhancing triplex stability, we synthesized daunomycin-conjugated GT TFOs targeting this site in the *c-myc* gene promoter.

The TFO target sequence was within a highly conserved region known to regulate the activity of the P2 promoter, which is the major transcription initiation site in the *c-myc* gene [36, 37]. A purine-rich tract is present about 40 bp upstream of the P2 start site [36, 37]. This region overlaps with the binding sites for transcription factors, such as Sp1/Sp3, MAZ, Ets, E2F and Stat 3 and is absolutely required for transcription of the *c-myc* gene [39, 40]. TFOs directed to the purine-rich core sequence were found to inhibit binding of transcription factors, transcription in vitro and promoter activity in cells [34, 35]. An antiparallel 23-mer GT TFO that we designed to target this site inhibited *c-myc* expression and cell proliferation [36, 37]. Therefore, we synthesized oligonucleotides conjugated to daunomycin to target the site near the P2 promoter (Fig. 5).

An 11-mer GT TFO, **dauno-GT11A**, was directed to the 3′ portion of the poly-purine sequence. We had previously shown that the 11-bp homo-purine sequence (GGAGGGAGGGA) was sufficient to drive triplex DNA formation in the antiparallel orientation [38]. However, unlike the 23-mer GT TFO, the unmodified 11-mer TFO formed a complex that was unstable at physiological and near-physiological temperatures ($T_m \leq 37\,°C$), thereby offering an ideal system for testing the effects of daunomycin on triplex stability [38]. Therefore, most of the initial studies were done with this 11-bp daunomycin-conjugated GT TFO **dauno-GT11A**. A second TFO, **dauno-GT11B**, was directed to a similarly short 11-bp purine-rich sequence (GGGAAGGGAGA) located about 100 bp downstream of the P2 transcription start site (Fig. 5). This second target site did not overlap with any known transcription factor binding site. We hypothesized that binding of **dauno-GT11B** at this site could interfere with the assembly of the transcription initiation complex or block transcription elongation. All the TFOs were synthesized with a phosphodiester backbone and covalently linked via a hexamethylene bridge to the O-4 position on the ring D of daunomycin. TFOs were also modified at the 3′ end by addition of a propanediol tail to increase nuclease resistance [41].

Two additional oligonucleotides, **dauno-CO11** and **dauno-GT11C**, were used as controls in these studies (Fig. 5). **Dauno-CO11** matched in parallel orientation a sequence adjacent to the **dauno-GT11A** binding sequence [38]. **Dauno-GT11C** was identical in sequence to **dauno-GT11A** but was linked to the amino group of daunomycin in an incorrect position through an amide bond (see compound **23** in Table 2).

At the time we started working on this project, the previous study with daunomycin-conjugated CT TFOs showing a considerable increase in binding affinity of the conjugated compared to the unconjugated TFO had been particularly encouraging [12]. However, it remained to be seen whether this approach would work at physiological pH and temperature and, ultimately, whether it would lead to increased activity in cells. Using the experimental tools available in our laboratory (i.e., *c-myc* promoter reporter vectors and cell types with distinct *c-myc* expression levels), we set out to address these questions regarding the activity and potential uses of daunomycin-conjugated TFOs as bioactive molecules and therapeutic compounds: Does conjugation to daunomycin improve DNA binding and biological activity of TFOs? Does it interfere with sequence-selectivity and does it hinder target-selectivity of the conjugated TFOs in cells?

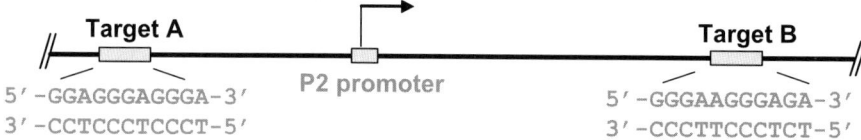

Dauno-GT11A	Dauno-$^{5'}$TGGGTGGGTGG$^{3'}$---
Dauno-GT11B	Dauno-$^{5'}$TGTGGGTTGGG$^{3'}$---
Dauno-CO11	Dauno-$^{5'}$GGGTTTTTGTT$^{3'}$---
Dauno-GT11C	NH-Dauno-$^{5'}$TGGGTGGGTGG$^{3'}$---

Fig. 5 Triplex target sites in the *c-myc* gene promoter and daunomycin-conjugated TFOs. *Top*, Target A and B are two 11-bp sequences located 40 bp upstream and 100 bp downstream of the P2 promoter, respectively. *Bottom*, dauno-TFOs (dauno-GT11A and **dauno-GT11B**) and control oligonucleotides (**dauno-CO11** and **dauno-GT11C**) used in the studies. NH-Dauno symbolizes that the in this conjugate the dauno moiety is linked through the amino-group (cfr. comp. 23 in Table 2), --- symbolizes the propandiol tail at the 3′ end of the conjugate

4.2.1
Binding of Daunomycin-conjugated TFOs to Double-Stranded DNA

Gel mobility shift assays showed that binding of **dauno-GT11** was efficient with approximately 50% of triplex DNA seen at a concentration of 250 nM of dauno-TFO (Fig. 6A). The affinity of the 11-mer dauno-TFO for the target sequence was in fact comparable or even greater than that of longer unmodified TFOs directed to the same site [34–37]. Triplex DNA formation was not detected in samples incubated with the unmodified **GT11**, which dissociated from the duplex when gel electrophoresis was done at temperatures $\geq 20\,^{\circ}\mathrm{C}$ (Fig. 6B). Thus, the addition of daunomycin clearly increased the triplex stability and allowed its persistence at physiological temperatures. Similar results were obtained with dauno-GT11B that bound with high affinity to its own target duplex [42]. The control oligonucleotide **dauno-CO11** did not bind to the target duplex (Fig. 6C).

To confirm that daunomycin attached to the TFOs intercalated at the duplex–triplex junction and its intercalation was important for triplex stabilization, we performed additional assays using a shorter duplex lacking five nucleotides at the 3' of the TFO binding site. The shortening of the duplex at the end of the triplex binding site would not affect binding of the TFO

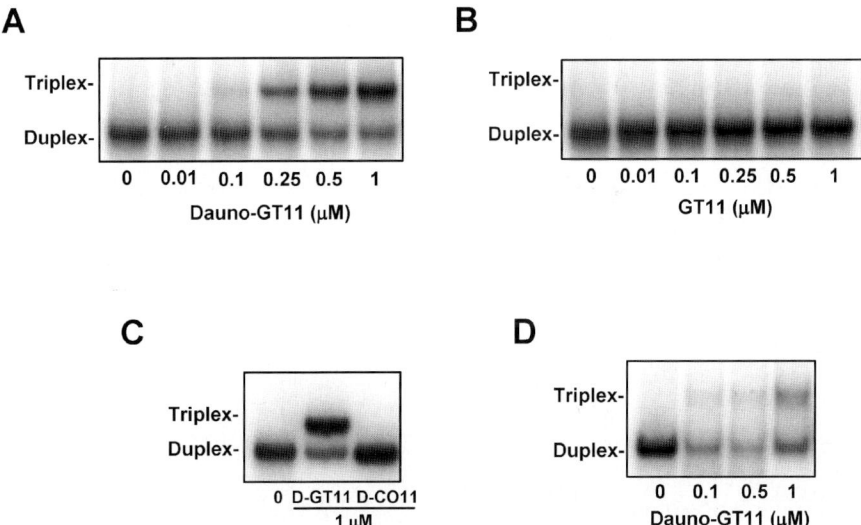

Fig. 6 Triplex DNA formation by daunomycin-conjugated TFOs. Oligonucleotides corresponding to the different duplex targets were incubated with dauno-TFO (**dauno-GT11A**) or control oligonucleotides (**dauno-CO11** or **dauno-GT11C**) and triplex formation assessed by gel mobility shift assay. **A** Binding of **dauno-GT11A** to the 28-bp duplex. **B** Binding of non-conjugated GT11 to the 28-bp duplex. **C** Binding of **dauno-GT11A** and **dauno-CO11** to the 40-bp duplex. **D** Binding of **dauno-GT11A** to the shorter 23-bp duplex

but would hinder the intercalation of daunomycin. Indeed, binding of **dauno-GT11A** to the shortened duplex was considerably reduced, almost to the level seen with the unmodified TFO (Fig. 6D). Thus, the increased stability of the triplex formed by **dauno-GT11A** depended upon the intercalation of daunomycin at a site near the duplex–triplex junction.

To further prove that the attached daunomycin did not drive per se the binding of dauno-TFOs but only secondarily stabilized the triplex DNA complex, we synthesized an oligonucleotide, **dauno-GT11C**, with the sequence identical to dauno-GT11A but conjugated to the amino sugar instead of the aglycone of daunomycin. This mode of attachment would eliminate any contribution of daunomycin to the triplex stability upon binding of the oligonucleotide to the duplex [12] (cf. compounds **22** and **23** in Table 2). Gel-shift assays confirmed our prediction showing that **dauno-GT11C** was unable to form a stable triplex, like to the non-conjugated **GT11** [43].

Overall, the in vitro binding assays showed that there was a considerable increase in triplex stability upon conjugation of antiparallel TFOs with daunomycin. This was consistent with the observations made previously with parallel CT TFOs [12] Thus, daunomycin stabilized binding of TFOs both in the parallel and antiparallel orientation. The potency of daunomycin as a DNA-binding agent and the high degree of stability conferred to triplex DNA upon conjugation to a TFO could be explained by its mode of interaction with double-stranded DNA. Both the anthraquinone and amino sugar make contact with the DNA at the intercalation site and the attachment of the TFO to position O-4 of the anthraquinone preserved the orientation of the intercalating moiety in the double helix, while maintaining the TFO and the amino sugar in their correct positions in the major and minor groove, respectively. Indeed, modifications of the site of attachment, as we did with the **dauno-GT11C**, eliminated completely any triplex stabilizing effect of the anthracycline.

4.2.2
Sequence-specific Recognition of the Target Sequence by the Daunomycin-conjugated TFOs

Attachment of a non-sequence-specific DNA-binding agent could affect the specificity of the interaction of the TFO with duplex DNA. In a previous study, random intercalation of daunomycin-conjugated TFOs was excluded by fluorescence-quenching experiments [12]. Intercalation and quenching of the chromophore occurred only when the DNA contained the appropriate target sequence, thus proving that site recognition and binding were dictated exclusively by the oligonucleotide component of the intercalator-conjugated TFO [12]. Although the previous studies had suggested a specific interaction of dauno-GT TFOs with their duplex targets, it was critical to provide more direct evidence that their binding was sequence-specific. We addressed this issue by performing gel mobility shift assays and DMS footprinting studies to show that dauno-TFOs were able to bind exclusively to their target sequences in the *c-myc* gene promoter.

We examined first triplex DNA formation by **dauno-GT11A** on a mutated target duplex by gel mobility shift assays. The mutated duplex contained four mismatched bases in the 11-bp target sequence. These centrally located mismatches were expected to destabilize binding of the oligonucleotide without affecting the intercalation of daunomycin. **Dauno-GT11A** was unable to bind to the mutated duplex, indicating that binding required the presence of the correct target sequence in the duplex [43]. Similarly, **dauno-GT11B** bound efficiently to the target duplex (target B), but did not bind to the noncomplementary target A, even though the two sequences were quite similar [42]. Thus, despite the presence of daunomycin, binding of dauno-TFOs was driven by the oligonucleotide sequence and required perfect matching with the sequence in the duplex.

DMS footprinting was used to examine further the specificity of the interaction of dauno-TFOs with duplex DNA. DMS footprinting is a more stringent test of the ability of TFOs to recognize unique sequences in DNA than gel mobility shift assays. In our study, a 339-bp fragment of the *c-myc* promoter was incubated with **dauno-GT11A** or **dauno-CO11** and then analyzed on sequencing gels [43]. Triplex DNA formation results in an area of protection ("footprint") from DMS/piperidine cleavage at the TFO binding site. Binding of **dauno-GT11A** was clearly visible as an area of protection corresponding exactly to the 11-bp target sequence within the larger purine-rich region of the *c-myc* gene promoter (Fig. 7).

Instead, no sites of protection were seen in samples incubated with **dauno-CO11** (Fig. 7). For comparison, we show the footprint formed by the 23-mer GCT-TFO that was complementary to the entire 23-bp *c-myc* target sequence [38]. Binding of a 23-mer GCT-TFO produced a larger area of protection corresponding to the entire purine-rich sequence in the c-myc promoter and including the 11-bp target site of **dauno-GT11A** (Fig. 7).

The extent of the protection achieved with **dauno-GT11A** was determined by densitometric analysis. Interestingly, more than 90% of the target sequence was protected in the presence of **dauno-GT11A**, confirming the high binding affinity of the dauno-TFO. The unconjugated **GT11** gave a much weaker footprint at the same concentrations, consistent with reduced triplex stability [38]. The interaction of dauno-TFOs with duplex DNA was highly sequence-specific, since no other region of the 339-bp fragment showed any alteration of the DMS cleavage pattern upon incubation with **dauno-GT11A**. This is remarkable since the *c-myc* promoter fragment used for footprinting contained multiple sequences (e.g., GAACGGAGGGA and AGAGGGAGCGA), which were very similar to the target site and could have allowed binding of the dauno-TFO.

1 2 3 4 5 6   7 8

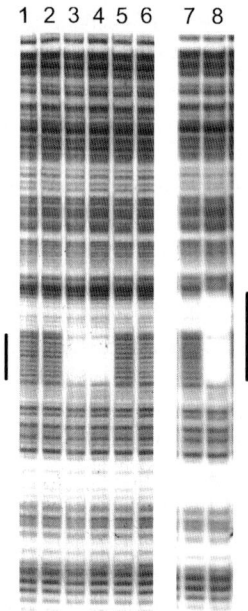

Fig. 7 Sequence-specific binding by the daunomycin-conjugated TFOs. A radiolabeled 339-bp fragment of the *c-myc* promoter was incubated with oligonucleotides and then treated with DMS and piperidine to cleave unprotected guanines. Samples were analyzed on sequencing gels under denaturing conditions. *Bars* indicate the position of the 11-bp and 23-bp target sequences. *Lane 1*, no oligonucleotide control; *lane 2*, 0.1 μM **dauno-GT11A**; *lane 3*, 1 μM **dauno-GT11A**; *lane 4*, 10 μM **dauno-GT11A**; *lane 5*, 1 μM **dauno-CO11**; *lane 6*, 10 μM **dauno-CO11**; *lane 7*, 5 μM **CT11** control oligonucleotide; *lane 8*, 5 μM 23-mer **GTC** TFO

Taken together, these studies provided compelling evidence that binding of dauno-TFOs depended strictly on the oligonucleotide sequence and was limited to the specific target site. Gel-shift assays confirmed the inability of dauno-TFOs to bind to duplex DNA with altered sequences. Footprinting assays showed that dauno-TFOs bound with high affinity to the target site but were able to discriminate against non-target sequences that contained even a limited number of mismatched bases. The presence of the TFO probably abolished random binding of daunomycin to DNA. Repulsion between the oligonucleotide tail attached to the daunomycin and duplex DNA that did not contain the complementary sequence prevented DNA intercalation at non-target sites. Therefore, binding of the intercalator-conjugated TFOs took place only at sites where the TFOs could find the exact matching sequence.

4.2.3
Transcriptional Inhibition by Daunomycin-Conjugated TFOs

In vitro transcription assays confirmed that triplex formation directed by dauno-TFOs blocked transcription from the P2 promoter site [43]. The block of transcription initiation by **dauno-GT11A** was probably due to its ability to prevent binding of transcription factors, like Sp1, to the site near the P2 promoter [38]. Interestingly, **dauno-GT11A** did not have any effect on Sp1 binding to a similar sequence upstream of the P1 promoter. The next logical step was to determine whether the dauno-TFOs were able to affect *c-myc* promoter activity and gene transcription in cells. This was done initially using a luciferase reporter system, in which the reporter gene (i.e., firefly luciferase) was put under the control of the *c-myc* promoter. Reporter assays are a reliable and rapid way to assess transcription inhibitory activity of small molecules and oligonucleotides in cells [44, 45]. In these studies, cells were transfected with the reporter plasmids along with dauno-TFOs (Fig. 8).

After 24 h, luciferase activity was measured in cell extracts. **Dauno-GT11A** induced a dose-dependent inhibition of the *c-myc* reporter activity with 70–80% reduction at doses ranging from 0.25 to 1 μM [43]. Similar results were obtained with distinct reporter constructs and in different cell types (e.g., prostate and breast cancer cells). **Dauno-GT11B** also inhibited *c-myc* promoter activity, while the control oligonucleotide **dauno-CO11** did not [42, 43]. This was consistent with the fact that the effects of dauno-TFOs on the *c-myc* promoter were sequence-specific and not due to delivery of daunomycin into cells.

To further rule out non-triplex mediated mechanisms of inhibition of *c-myc* promoter activity, similar experiments were performed with **dauno-GT11C** that was linked to daunomycin via the amino sugar and unable to bind to the target. **Dauno-GT11C** was unable to inhibit *c-myc* promoter activity in cells, consistent with its inability to form triplex DNA [46]. Taken together, these results confirmed that inhibition of *c-myc* promoter activity by dauno-TFOs required triplex formation at the target site and ruled out alternative non-triplex dependent mechanisms that might lead to non-specific activity.

These data suggested that dauno-TFOs could form stable triple helical complexes with the target sequences and inhibit transcription of the *c-myc* gene in cells. To investigate this aspect further, dauno-TFOs were delivered into cells using cationic lipid preparations, like DOTAP. Like unmodified oligonucleotides, cellular and nuclear uptake of dauno-TFOs was enhanced by co-administration with cationic lipids [42, 46]. After 24 h from transfection, the expression of *c-myc* was determined by RT-PCR and Western blotting. *c-myc* RNA was reduced in cells treated with 1 μM of either **dauno-GT11A** or **dauno-GT11B** compared to untreated and control oligonucleotide treated cells (Fig. 8). The *c-myc* protein level was similarly reduced in cells

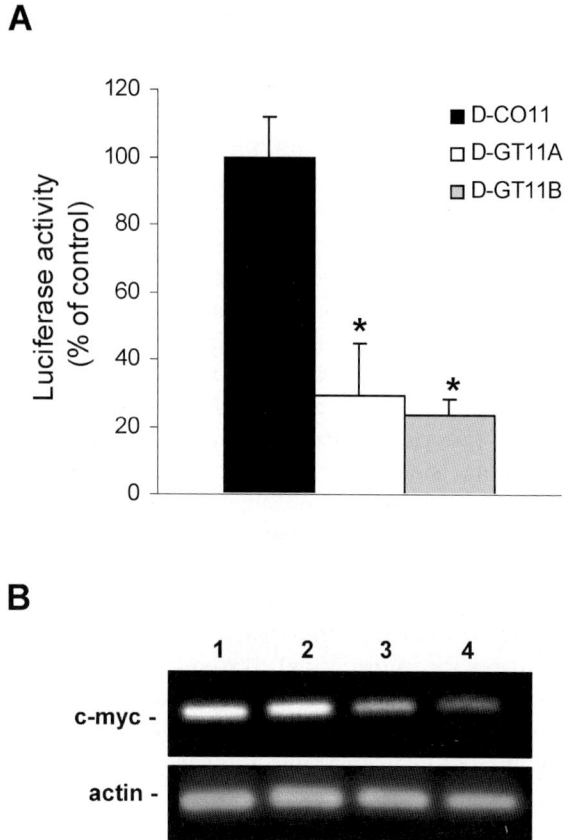

Fig. 8 Inhibition of *c-myc* transcription by daunomycin-conjugated TFOs. **A** Cells were transfected with the *c-myc* reporter, a control vector and 1 μM of dauno-TFOs or control oligonucleotide. Luciferase activity was measured after 24 h. * P < 0.05 compared to control cells. **B** Cells were transfected without oligonucleotide (*lane 1*) or with 1 μM of **dauno-GT11C** (*lane 2*), **dauno-GT11A** (*lane 3*) and **dauno-GT11B** (*lane 4*). Total RNA was extracted after 24 h and *c-myc* transcript levels were determined by RT–PCR

treated with dauno-TFOs compared to control cells [42, 43]. Interestingly, dauno-TFOs did not affect *c-myc* RNA and protein level in cells transiently transfected with a *c-myc* expression vector and expressing the gene from a heterologous promoter lacking the TFO target sequence [42].

We also monitored the expression of two genes, *ets* and *ets-2*, which had sequences with great similarity to the TFO target sequence in their regulatory regions [46]. The *ets-2* gene had a sequence (GGA<u>A</u>GGAGGGA) with a single mismatch compared to the *c-myc* site and corresponding to a critical Sp1 binding site [45]. The *ets-1* gene had a similar sequence (<u>A</u>GAGGGAGGGA) with a single mismatched nucleotide and adjacent to a Sp1 site [47]. Expression of both genes was not affected in cells transfected with dauno-

TFOs, indicating that their antigene activity was selective for the target gene [42, 43].

These data provided evidence of the ability of dauno-TFOs to reach the chromosomal target site, form stable triplex DNA under physiological conditions and affect transcription in cells. The experiments with **dauno-GT11C**, a similar daunomycin-conjugated oligonucleotide unable to form triplex DNA, demonstrated that the effects of dauno-TFOs on transcription could not be attributed to a decoy-like mechanism (e.g., trapping of transcription activators required for *c-myc* promoter activity). Indeed, multiple lines of evidence support the conclusion that the activity of dauno-TFOs was sequence-specific and consistent with a triplex-mediated mechanism. The interaction of dauno-TFOs with the respective targets in vitro was strictly sequence-dependent as shown by the inability to bind to alternative target sites despite close sequence homology [42, 43]. Inhibition of transcription factor binding was also sequence-dependent, excluding both binding of the TFOs to non-target sequences and direct interaction of TFOs with transcriptional activators via a decoy or aptamer effect. However, we cannot exclude that additional mechanisms are involved in the inhibition of *c-myc* transcription by dauno-TFOs. The presence of the intercalating agent might induce DNA damage or recruit DNA repair enzymes with secondary effects on gene transcription. Another possibility due to the presence of the daunomycin moiety is the recruitment of DNA topoisomerase II that might result in topoisomerase-mediated DNA cleavage. Conjugation of topoisomerase-interactive compounds to TFOs has been shown to direct the DNA cleaving activity of topoisomerases to specific genomic sites [48, 49]. However, any of these additional mechanisms would still require triplex-directed binding of dauno-TFOs at the target sites in the *c-myc* promoter.

4.2.4
Anticancer Activity of Daunomycin-Conjugated TFOs

To assess the potential of dauno-TFOs for therapeutic applications, we investigated their effects on proliferation and survival of normal and cancer cells. The *c-myc* gene is over-expressed in many cancers, including prostate cancer [28, 30, 50]. Studies with transgenic mice indicate that *c-myc* has an important role in development of prostate cancer [51]. *C-myc* also contributes to the androgen-independent phenotype of advanced prostate cancer [52]. Reducing *c-myc* levels was shown to be sufficient to cause growth arrest and death of cultured cancer cells and tumor regression in mice [30]. Thus, dauno-TFOs directed to the *c-myc* promoter might inhibit proliferation of prostate cancer cells that depend on constitutive expression of the gene. To test this hypothesis, we evaluated the effects of the *myc*-targeting dauno-TFOs on growth and survival of established prostate cancer cell lines. Cells were transfected with dauno-TFOs and cell viability was measured after

4 days using a standard colorimetric assay. Both **dauno-GT11A** and **dauno-GT11B** inhibited growth of prostate cancer cells (e.g., DU145, PC3, LNCaP and 22Rv1) by $\geq 50\%$ at concentrations of $0.5-1\,\mu M$ (Fig. 9).

Control oligonucleotides **dauno-CO11** and **dauno-GT11C** had minimal effects on cell growth with no significant differences with respect to untreated control cells [42]. Dauno-TFOs inhibited proliferation of prostate cancer cells also in long-term clonogenic assays [42]. The number of colonies was significantly lower in cells treated with dauno-TFOs compared to cells treated with vehicle or a control oligonucleotide ($p < 0.05$). Since daunomycin is a potent cytotoxic drug, one must be careful in interpreting the results of cell growth inhibition assays with dauno-TFOs. However, our data strongly argued against the possibility that the anti-proliferative effects of dauno-TFOs were due to toxicity associated to the daunomycin moiety. When daunomycin was conjugated to oligonucleotides unable to form triplex DNA either because of the nucleotide sequence (**dauno-CO11**) or the attachment mode (**dauno-GT11C**), there was no evident cytotoxicity both in short- and long-term assays [42]. This was consistent with the fact that conjugation to oligonucleotides altered many biophysical and biochemical properties of daunomycin, including cellular uptake, intracellular trafficking, and DNA binding [43].

Dauno-TFOs inhibited growth of prostate cancer cells constitutively expressing high levels of *c-myc*. We expected that on the contrary, selective down-regulation of *c-myc* would be less toxic to cells that express low levels of the target gene, like normal human fibroblasts. This would also rule out the concern that dauno-TFOs, because of their short length and the presence

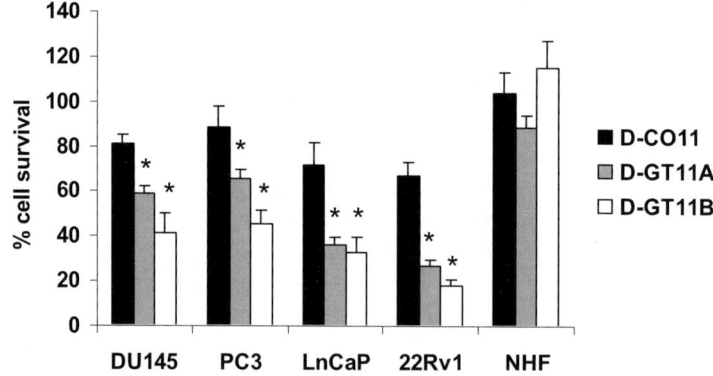

Fig. 9 Antiproliferative activity of daunomycin-conjugated TFOs in prostate cancer cells and normal human fibroblasts. Prostate cancer cells (DU145, PC3, LNCaP, and 22Rv1) and normal human fibroblasts (NHF) were transfected with $0.5\,\mu M$ of **dauno-GT11A**, **dauno-GT11B** or **dauno-CO11**. Percent of cell survival (means $\pm$ SD) compared to untreated control cells was determined 4 days later using a colorimetric assay. * $p < 0.01$ compared to control cells

of daunomycin, would bind to multiple sites in the genome with identical or similar sequences producing non-target specific cytotoxic effects. If this were the case, then even low-expression cells would be affected. To test this hypothesis, primary cultures of normal human fibroblasts were transfected with dauno-TFOs and cell growth and viability was measured by a colorimetric assay. Under these conditions, normal fibroblasts took up dauno-TFOs with an intracellular distribution similar to that of prostate cancer cells [42]. The total amount of intracellular dauno-TFOs measured by flow cytometry was similar in normal fibroblasts and prostate cancer cells and in both cell types dauno-TFOs accumulated both in the cytoplasm and nucleus [42]. Furthermore, dauno-TFOs inhibited *c-myc* promoter activity and endogenous *c-myc* expression in normal fibroblasts as in prostate cancer cells [42]. However, despite the ability to inhibit *c-myc* transcription, dauno-TFOs did not have any effect on the growth of normal fibroblasts (Fig. 9). Thus, *c-myc*-targeting dauno-TFOs had antiproliferative effects only toward cells with high *c-myc* levels, while they were minimally toxic against normal cells with low expression of the gene. This was in striking contrast to the effects of free daunomycin that was toxic both in cancer and normal cells independently of the *c-myc* expression level [43]. These suggested that the antiproliferative activity of dauno-TFOs was a direct consequence of their ability to interfere with *c-myc* transcription.

5
Conclusions

Thus far, our studies have collectively provided evidence of the activity of dauno-TFOs as transcriptional repressors in vitro and in cells and may open new avenues for design of selective gene-targeted therapeutics that might be particularly useful for cancer treatment. The activity of dauno-TFOs was consistent with a triplex-mediated mechanism and was clearly different from the non-gene selective effects of free daunomycin, supporting the idea that daunomycin and dauno-TFOs have distinct modes of action. DNA binding, transcriptional inhibition, and antiproliferative activity of dauno-TFOs appeared to be dictated exclusively by the oligonucleotide component of the intercalator conjugates.

Our studies also revealed that triplex-mediated targeting of relatively short homo-purine sequences in genomic DNA and transcriptional modulation of gene expression are possible with the addition of a strong intercalating and triplex stabilizing moiety, like daunomycin, to the TFOs. We believe that the range of applications of the triplex-mediated gene targeting strategy could be considerably increased by this approach.

Acknowledgements Work in the authors' laboratories has been supported by grants from: Associazione Italiana per la Ricerca sul Cancro, Swiss Cancer League, Fondazione Ticinese per la Ricerca sul Cancro, National Cancer Institute, US Army Breast Cancer Research Program, Menarini Ricerche Spa and M.U.R.S.T. - Italy. The authors wish to thank Federico Arcamone (ISOF-CNR, Bologna, Italy), Anna Garbesi (ISOF-CNR, Bologna, Italy), and Giuseppina Carbone (IOSI, Bellinzona, Switzerland) for their help and advice, and all coworkers that have contributed to this work.

References

1. Fenselfed G, Davies DR, Rich A (1957) J Am Chem Soc 79:2023
2. Watson JD, Crick FHC (1953) Nature 171:737
3. Moser HE, Dervan PB (1987) Science 238:645
4. Francois JC, Saison-Behmoaras T, Chassignol M, Thuong NT, Hélène C (1988) CR Acad Sci III 307:849
5. Hélène C (1991) Anti-Cancer Drug Des 6:569
6. Goni JR, de la Cruz X, Orozco M (2004) Nucleic Acids Res 32:354
7. Thuong NT, Hélène C (1993) Angew Chem Int Ed 32:666
8. Giovannangeli C, Diviacco S, Labrousse V, Gryaznov S, Charneau P, Hélène C (1997) Proc Natl Acad Sci USA 94:79
9. Macris MA, Glazer PM (2003) J Biol Chem 278:3357
10. Asseline U, Bonfils E, Kurfurst R, Chassignol M, Roig V, Thuong NT (1992) Tetrahedron 48:1233
11. Cirilli M, Bachechi F, Ughetto G, Colonna FP, Capobianco ML (1993) J Mol Biol 230:878
12. Garbesi A, Bonazzi S, Zanella S, Capobianco ML, Giannini G, Arcamone F (1997) Nucleic Acids Res 25:2121
13. Bernardi L, Masi P, Sapini O, Suarato A, Arcamone F (1979) Farmaco 34:884
14. Capobianco ML, De Champdore M, Arcamone F, Garbesi A, Guianvarc'h D, Arimondo PB (2005) Bioorg Med Chem 13:3209
15. Israel M, Taube D, Seshadri R, Idriss JM (1986) J Med Chem 29:1273
16. Uhlmann E, Engels J (1986) Tetrahedron Lett 27:1023
17. Iyer RP, Egan W, Regan JB, Beaucage SL (1990) J Am Chem Soc 112:1253
18. Guzaev A, Lonnberg H (1997) Tetrahedron Lett 38:3989
19. Cheruvallath ZS, Carty RL, Moore MN, Capaldi DC, Krotz AH, Wheeler PD, Turney BJ, Craig SR, Gaus HJ, Scozzari AN, Cole DL, Ravikumar VT (2000) Organic Process Res Dev 4:199
20. Kumar P, Gupta KC, Rosch R, Seliger H (1997) Chem Lett 26:1231
21. Guzaev A, Lonnberg H (1997) Tetrahedron Lett 38:3989
22. Borer PN (1975) Optical properties of nucleic acids. Absorption and circular dichroism spectra. In: Fasman GD (ed) Handbook of Biochemistry and Molecular Biology. CRC Press, Boca Raton, Florida, p 589
23. Chaires JB (1983) Biochem 22:4204
24. Xodo LE, Manzini G, Ruggiero J, Quadrifoglio F (1988) Biopolymers 27:1839
25. Giovannangeli C, Rougée M, Garestier T, Thuong NT, Hélène C (1992) Proc Natl Acad Sci USA 89:8631
26. Capobianco ML, De Champdoré M, Francini L, Lena S, Garbesi A, Arcamone F (2001) Bioconjug Chem 12:523

27. Giovannangeli C, Perrouault L, Escudé C, Thuong N, Hélène C (1996) Biochem 35:10539
28. Dang CV (1999) Mol Cell Biol 19:1
29. Grandori C, Cowley SM, James LP, Eisenman RN (2000) Annu Rev Cell Dev Biol 16:653
30. Pelengaris S, Khan M, Evan G (2002) Nat Rev Cancer 2:764
31. Cooney M, Czernuszewicz G, Postel EH, Flint SJ, Hogan ME (1988) Science 241:456
32. Postel EH, Flint SJ, Kessler DJ, Hogan ME (1991) Proc Natl Acad Sci 88:8227
33. Thomas TJ, Faaland CA, Gallo MA, Thomas T (1995) Nucleic Acids Res 23:3594
34. Kim HG, Miller DM (1995) Biochemistry 34:8165
35. Kim HG, Reddoch JF, Mayfield C, Ebbinghaus S, Vigneswaran N, Thomas S, Jones DE Jr, Miller DM (1998) Biochemistry 37:2299
36. Catapano CV, McGuffie EM, Pacheco D, Carbone GM (2000) Biochemistry 39:5126
37. McGuffie EM, Pacheco D, Carbone GM, Catapano CV (2000) Cancer Res 60:3790
38. McGuffie EM, Catapano CV (2002) Nucleic Acids Res 30:2701
39. Majello B, De Luca P, Suske G, Lania L (1995) Oncogene 10:1841
40. Albert T, Wells J, Funk JO, Pullner A, Raschke EE, Stelzer G, Meisterernst M, Farnham PJ, Eick D (2001) J Biol Chem 276:20482
41. Pandolfi D, Rauzi F, Capobianco ML (1999) Nucleosides Nucleotides 18:2051
42. Napoli S, Negri U, Arcamone F, Capobianco ML, Carbone GM, Catapano CV (2006) Nucleic Acids Res 34:734
43. Carbone GM, McGuffie E, Napoli S, Flanagan CE, Dembech C, Negri U, Arcamone F, Capobianco ML, Catapano CV (2004) Nucleic Acids Res 32:2396
44. Albertini V, Jain A, Vignati S, Napoli S, Rinaldi A, Kwee I, Nur-e-Alam M, Bergant J, Bertoni F, Carbone GM, Rohr J, Catapano CV (2006) Nucleic Acids Res 34:1721
45. Carbone GM, McGuffie EM, Collier A, Catapano CV (2003) Nucleic Acids Res 31:833
46. Carbone GM, Napoli S, Valentini A, Cavalli F, Watson DK, Catapano CV (2004) Nucleic Acids Res 32:4358
47. Jorcyk CL, Watson DK, Mavrothalassitis GJ, Papas TS (1991) Oncogene 6:523
48. Arimondo P, Bailly C, Boutorine A, Asseline U, Sun JS, Garestier T, Hélène C (2000) Nucleosides Nucleotides Nucleic Acids 19:1205
49. Arimondo PB, Boutorine A, Baldeyrou B, Bailly C, Kuwahara M, Hecht SM, Sun JS, Garestier T, Hélène C (2002) J Biol Chem 277:3132
50. Spencer CA, Groudine M (1991) Adv Cancer Res 56:1
51. Ellwood-Yen K, Graeber TG, Wongvipat J, Iruela-Arispe ML, Zhang J, Matusik R, Thomas GV, Sawyers CL (2003) Cancer Cell 4:223
52. Bernard D, Pourtier-Manzanedo A, Gil J, Beach DH (2003) J Clin Invest 112:1724

Top Curr Chem (2008) 283: 73–97
DOI 10.1007/128_2007_5
© Springer-Verlag Berlin Heidelberg
Published online: 21 November 2007

Acid-Sensitive Prodrugs of Doxorubicin

Felix Kratz

Tumor Biology Center, Macromolecular Prodrugs, Breisacher Straße 117,
79106 Freiburg im Breisgau, Germany
felix@tumorbio.uni-freiburg.de

Abstract The endosomal and/or lysosomal pathway of macromolecules, as well as the slightly acidic extracellular environment in solid tumors, form the rationale for designing carrier-linked prodrugs with pH-dependent linkers. In the past 20 years, a spectrum of acid-sensitive doxorubicin prodrugs has been developed with antibodies, serum proteins, and synthetic polymers. For a number of these, a convincing proof of concept has been obtained preclinically, showing an enhanced therapeutic efficacy of the prodrugs compared to free doxorubicin in tumor models. Clinically, the (6-maleimidocaproyl)-hydrazone derivative of doxorubicin, which binds either to the monoclonal antibody BR96 or to endogenous albumin, has been evaluated in clinical trials.

Keywords Acid-labile · Acid-sensitive · Doxorubicin · Drug conjugates · Prodrugs

1
Introduction

In the field of cancer chemotherapy, designing and developing truly tumor-specific prodrugs remains a challenge. On one hand, active targeting strategies aim to exploit membrane-associated receptors or antigens for drug delivery. On the other hand, the enhanced vascular permeability and retention of macromolecules in the tumor tissue substantiates the concept of passive targeting. Consequently, research efforts have concentrated on conjugation of anti-cancer agents with a wide spectrum of carriers including antibodies, peptides, serum proteins, and synthetic polymers [1–3]. Conversely, low-molecular weight prodrugs of anti-cancer agents have been developed that do not bear an active or passive targeting moiety, but are activated by tumor-associated enzymes at the tumor site [4].

Anthracyclines, such as doxorubicin, epirubicin, idarubicin, and daunorubicin, are widely used to treat solid and hematological tumors. However, the clinical application of anthracyclines is limited by their dose-related side effects, which include bone marrow toxicity, gastrointestinal disorders, stomatitis, alopecia, acute and cumulative cardiotoxicity, as well as extravasation [5]. Bone marrow suppression is generally a dose-limiting toxicity. After each course of treatment, myelocytopenia and thrombocytopenia are the most prominent side effects, with their toxicity reaching its maximum 7 to 10 days after each treatment course, followed by a rapid recovery thereafter. Of special concern is the cumulative cardiotoxicity (cardiomyopathy and congestive heart failure) which is irreversible, and steadily increases once cumulative doses of doxorubicin and epirubicin exceed 500 and 900 mg/m^2, respectively [6].

Tumor-targeted delivery is based on the development of less toxic derivatives of the parent drug that are activated within the tumor or that carry an additional ligand with tumor-targeting properties, which transport the payload to the tumor, where the drug is then released, either intracellularly or extracellularly. Anthracyclines probably represent the most widely utilized class of anti-cancer agents used for the development of prodrugs [7].

The extracellular as well as intracellular drop in the pH value in solid tumors forms the rationale for the design of acid-sensitive prodrugs. This article summarizes various attempts to design prodrugs using the most prominent anthracycline, doxorubicin, with a focus on those derivatives that have reached an advanced preclinical or clinical stage.

2
Rationale for Developing Acid-Sensitive Prodrugs

The endosomal and/or lysosomal pathway of synthetic or natural polymers are attractive routes for the delivery of polymer-bound drugs to cells. Indeed, the significant drop in the pH-value, from 7.2–7.4 in the blood or extracellular spaces to 4.0–6.5 in the various intracellular compartments, is a fairly unique physical property in living systems. This property can be exploited for intracellular drug delivery by coupling drugs to macromolecular carriers through acid-sensitive bonds.

In general, macromolecules are taken up by cells either through receptor-mediated endocytosis, adsorptive endocytosis, or fluid-phase endocytosis [8–10]. Endocytosis is a complex process in which invaginations occur at the cell surface, and form endosomes, which migrate into the cytoplasm. Depending on the macromolecule and the kind of endocytotic process involved, a series of sorting steps take place. Endosomes are transported to certain cell organelles (e.g. the golgi apparatus), or they return to the cell surface (recycling), or they form primary and secondary lysosomes, respectively [9]. In the endosomes and lysosomes, a significant tumor drop in the pH-value takes place, from the physiological pH (7.2–7.4) in the extracellular space, to pH 6.5–5.0 in the endosomes, and to around pH 4.0 in primary and secondary lysosomes. Additionally, a great number of lysosomal enzymes,(such as phosphatases, nucleases, proteases, esterases, and lipases), become active in the acidic environment of these vesicles..

Furthermore, the microenvironment of tumors is reported to be slightly acidic in animal models and humans. Non-invasive techniques demonstrate that the pH-value in tumor tissue is often 0.5–1.0 units lower than in normal tissue [11]. This pH-shift could contribute to the extracellular release of drugs bound to the carriers through acid-sensitive linkers, especially if the drug is trapped by the tumor for longer periods of time. The acidic pH-value in solid tumors is, therefore, an intracellular and/or extracellular property, which can be exploited for the release of polymer-bound drugs into tumor cells.

3
Design of Acid-Sensitive Prodrugs with Doxorubicin

From a chemical point of view, doxorubicin is ideally suited for the design of prodrugs, due to the presence of two different functional groups, i.e. the 3'-amino group of the sugar moiety and the C-13-keto position. Acid-sensitive derivatives are developed by forming a carboxylic hydrazone bond at the C-13 carbonyl group, or by attaching a cis-aconityl spacer at the 3'-NH$_2$-group (see Fig. 1).

Both bonds are highly stable at pH values of 7.0–7.4. However, they release the anthracycline within a few hours at pH 5.

4
Doxorubicin Prodrugs with Antibodies

As of 1975, coupling of drugs to macromolecular carriers received an important impetus with the development of monoclonal antibodies due to the intellectual attractiveness of selectively targeting tumor-specific antigens or recep-

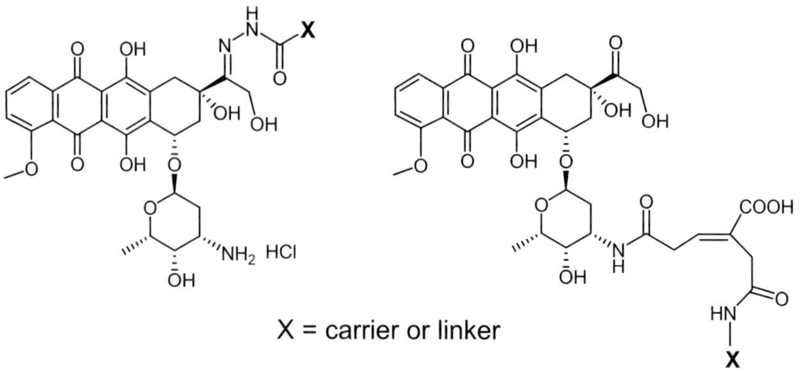

Fig. 1 Structure of acid-sensitive doxorubicin prodrugs with hydrazone linkers (*left*) or *cis*-aconityl linkers (*right*)

tors, ideally propagating the therapeutic concept of drug targeting founded on Paul Ehrlich's vision of "the magic bullet".

With respect to doxorubicin, the pharmaceutical research team at Bristol-Myers Squibb has developed monoclonal antibody conjugates with doxorubicin, which incorporate an acid-sensitive hydrazone bond, and a clinical candidate, the BR96-doxorubicin immunoconjugate, has been assessed in phase I and phase II studies (see Chap. 7).

In the late 1980s, researchers at Bristol-Myers Squibb synthesized a 6-maleimidodocaproyl and a 3-(2′-pyridinyldithio)propanoyl hydrazone derivative of doxorubicin (see Fig. 2) [12, 13].

Both compounds were coupled to thiol-bearing monoclonal antibodies, which bind to tumor-associated antigens with subsequent internalization of the antibody conjugate, allowing a release of doxorubicin in the acidic pH environment of endosomes and lysosomes. Such antibody conjugates have shown promising in vitro and in vivo activity [14–16].

Due to the high plasma stability of the resulting thioether bond formed after a reaction of the maleimide with thiol groups, the (6-maleimidocaproyl) hydrazone derivative of doxorubicin was selected for further clinical development in combination with the chimeric human/mouse monoclonal antibody BR96, which is specific for Lewis-Y, an antigen abundantly expressed on the surface of several human carcinomas. In this conjugate, known as the BR96-doxorubicin immunoconjugate, approximately eight molecules of doxorubicin are bound to the antibody. Therapy with the BR96-doxorubicin immunoconjugate induced complete remissions in a number of xenograft tumor models, and it was superior to unbound doxorubicin even at equivalent doxorubicin doses [14–16], which were below the effective dose for free doxorubicin. It also showed synergistic effects with paclitaxel in several xenograft models [17]. Phase I and phase II studies have been performed with this immunoconjugate (see Chap. 7) [18–20].

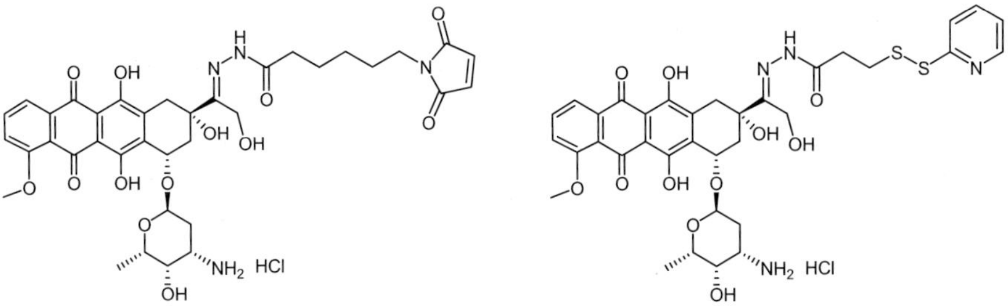

Fig. 2 Structure of (6-maleimidocaproyl) and 3-(2′-pyridinyldithio)propanoyl hydrazone derivative of doxorubicin, coupled to monoclonal antibodies

Interesting variants of the (6-maleimidocaproyl)hydrazone derivative of doxorubicin are branched systems, in which two doxorubicin molecules are linked to a maleimide spacer (see Fig. 3) [21, 22].

Applying these linkers achieved a 2-fold increase in the loading capacity (16 mol doxorubicin per mol mAb). These conjugates demonstrated an enhanced in vitro antigen-specific cytotoxicity, compared to the non-branched system [21].

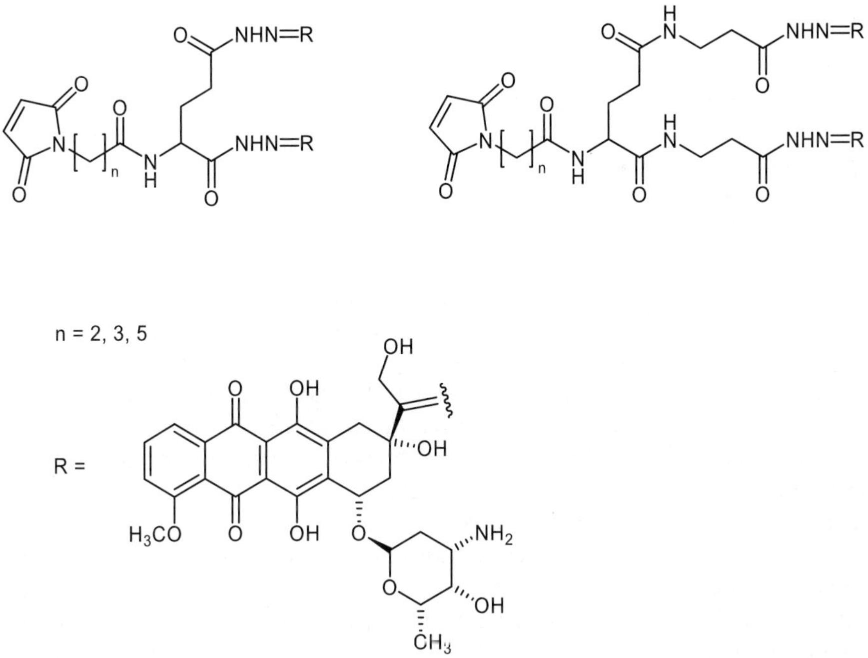

Fig. 3 Branched systems of (6-maleimidocaproyl) hydrazone derivatives of doxorubicin

5
Doxorubicin Prodrugs with Serum Proteins

Among blood proteins, serum albumin and transferrin have attracted the most interest, due to their potential as drug delivery systems for improving cancer chemotherapy. These proteins are suitable as drug carriers for a number of reasons: (a) They exhibit a preferential uptake in tumor tissue; (b) Tumor cells express high amounts of specific transferrin receptors on their cell surface; (c) They are readily available in a pure form exhibiting good biological stability; (d) They are biodegradable, non-toxic, and non-immunogenic.

There is meanwhile a large body of evidence available that demonstrates that albumin and transferrin accumulate in experimental solid tumors. Uptake of these macromolecules in tumor tissue has been demonstrated using various imaging methods, including labeling of the polymer with radioactive derivatives, fluorescent derivatives, or dye derivatives [3].

A number of doxorubicin conjugates with albumin and transferrin have been developed by Kratz and colleagues. In addition, acid-labile doxorubicin conjugates with lactosaminated albumin have been developed by Fiume et al. that target the asialoglycoprotein receptor over-expressed in liver cancer. The respective conjugates are described in detail below.

5.1
Doxorubicin Conjugates with Transferrin and Albumin

In the initial work by Kratz et al., maleimide derivatives of doxorubicin were synthesized during the first step, in which 3-maleimidobenzoic acid hydrazide or 4-maleimidophenylacetic acid hydrazide was bound to the 13-keto position of doxorubicin through an acid-sensitive carboxylic hydrazone bond. In the second step, the doxorubicin maleimide derivatives were coupled to thiolated albumin or transferrin, and the conjugates were isolated using size exclusion chromatography (see Fig. 4) [23, 24].

The results of subsequent in vitro studies with these conjugates showed in vitro efficacy comparable to free doxorubicin in 5–10 human tumor cell lines [25]. Interestingly, the corresponding acid-sensitive transferrin and albumin conjugates of doxorubicin demonstrated almost identical cytotoxicity. As shown by confocal laser scanning microscopy, there are marked differences between the intracellular distribution of the doxorubicin protein conjugates and unbound doxorubicin [26]. Free doxorubicin is initially localized in the cell nucleus and, with time, it is additionally observed in the golgi apparatus and mitochondria. Predominant sites of accumulation for doxorubicin, transferrin, and albumin conjugates are the golgi apparatus and mitochondria. Finally, the cellular distribution pattern and cytotoxicity are very similar for the acid-sensitive transferrin and albumin conjugates of doxorubicin.

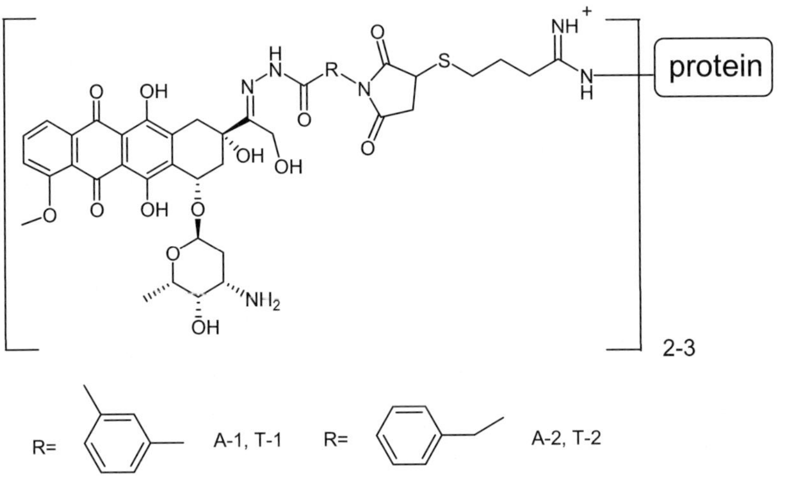

Fig. 4 Structures of transferrin and albumin conjugates with doxorubicin, containing an acid-sensitive carboxylic hydrazone bond

In vivo antitumor activity of acid-sensitive transferrin and albumin conjugates of doxorubicin (A-2 and T-2 – see Fig. 4) were evaluated against murine renal cell carcinoma (RENCA) and in the MDA-MB-435 mamma carcinoma [25, 27]. The maximum tolerated dose (MTD) of acid-sensitive doxorubicin transferrin and albumin conjugates was 2–3-fold higher than for free doxorubicin. The conjugates showed a significantly reduced toxicity (reduced lethality and body weight loss) with a concomitantly stable or improved antitumor activity, compared to the free drug. As observed in the in vitro analyses, there was no pronounced difference between identically constructed transferrin and albumin doxorubicin conjugates with regard to in vivo efficacy [25].

As a consequence, research efforts focused on the development of albumin drug conjugates, considering that the costs for obtaining albumin are 10-fold lower than for transferrin. In addition, we wanted to improve the coupling methods of drug derivatives, in order to obtain better defined drug albumin conjugates, which would have high purity, a constant drug loading ratio, and a minimal alteration of the three-dimensional protein structure.

Commercially available albumin is a mixture of mercaptalbumin and non-mercaptalbumin, containing approximately 20–60% free sulfhydryl groups per molecule albumin. This is due to the fact that the cysteine-34 position is blocked by sulfhydryl compounds, such as cysteine, homocysteine, or glutathione. Therefore, we developed a procedure for selective reduction of HSA with suitable agents, such as dithiothreitol (Cleland's reagent), in the first step, so that approximately one sulfhydryl group per molecule albumin can be determined (see Fig. 5).

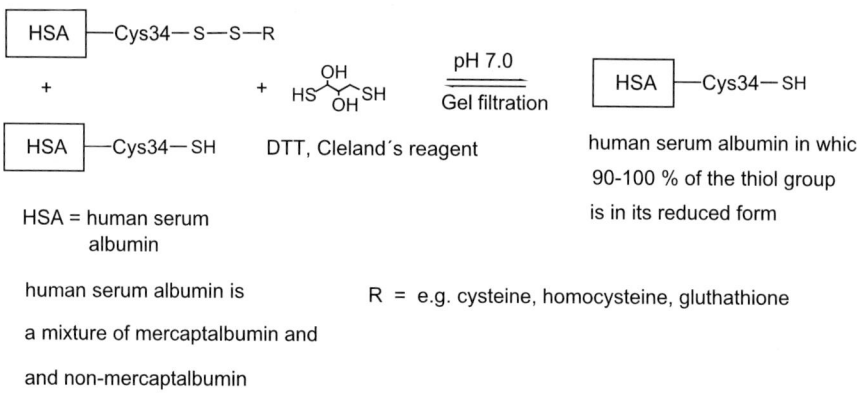

Fig. 5 Selective reduction of the cysteine-34 position of commercially available human serum albumin

In a second step, doxorubicin maleimide derivatives, such as the 4-male-imidophenylacetyl hydrazone derivative of doxorubicin (abbreviated DOXO-HYD), were coupled to this reduced form of albumin. The resulting conjugate, A-DOXO-HYD, was isolated through size-exclusion chromatography (see Fig. 6).

In the subsequent biological studies, the in vivo efficacy and pharmacokinetic properties of the acid-sensitive doxorubicin albumin conjugate, A-DOXO-HYD, were evaluated against murine metastatic renal cell carcinoma, (RENCA), in comparison to free doxorubicin at equitoxic dose [28]. At equitoxic dose, A-DOXO-HYD was superior compared to free doxorubicin against murine renal carcinoma. As shown in Fig. 7, therapy with the conjugate resulted in complete remissions of the primary kidney tumors at a dose of 4×12 mg/kg doxorubicin equivalents. Only two metastases in the lungs were observed. In contrast, mice treated with doxorubicin at the maximum tolerated dose of 4×6 mg/kg manifested clearly visible kidney tumors and a large number of lung metastases at the end of the experiment.

At a dose of 12 mg/kg doxorubicin equivalents, the AUC (0–72 h) was approximately two times higher for A-DOXO-HYD in the kidney tumor and in the liver. In contrast, the AUC (0–72 h) was approximately two times lower for A-DOXO-HYD in the healthy kidney and in the heart [28].

Encouraged by these results, we focused our work on a prodrug concept that uses *endogenous* albumin, the chief circulating protein in the blood stream (35–50 mg/mL), as a drug carrier [29, 30]. In this therapeutic strategy, the prodrug is designed to bind rapidly and selectively to the cysteine-34 position of circulating serum albumin after intravenous administration, thereby generating a macromolecular transport form of the drug in situ in the blood. We reasoned that exploiting circulating albumin as the drug carrier would have several advantages over ex vivo synthesized drug albumin conjugates,

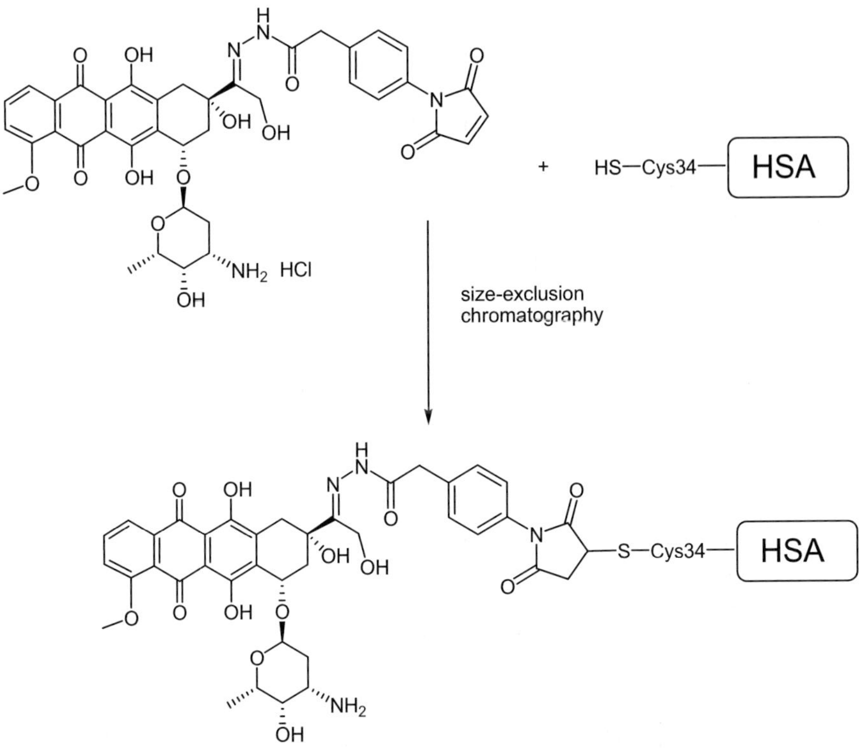

Fig. 6 Synthesis of albumin doxorubicin conjugate, A-DOXO-HYD, in which 4-maleimido-phenylacetyl hydrazone derivative of doxorubicin is bound to the cysteine-34 position of human serum albumin

due to the following: (a) The use of commercial and possibly pathogenic albumin is avoided; (b) Albumin-binding drugs are chemically well-defined and based on straight-forward organic chemistry; (c) Albumin-binding drugs are fairly simple and inexpensive to manufacture; (d) A broad range of drugs for developing albumin-binding drugs can be used; (e) The analytical requirements for defining the pharmaceutical products are comparable to any other low-molecular weight drug candidate.

The macromolecular prodrug approach targets the cysteine-34 position of albumin. Approximately 70 percent of circulating albumin in the blood stream is mercaptalbumin, containing an accessible cysteine-34, which is not blocked by endogenous sulfhydryl compounds, such as cysteine or glutathione. Considering that free thiol groups are not found on the majority of circulating serum proteins, except for albumin, cysteine-34 of endogenous albumin is a unique amino acid on the surface of the circulating protein.

A proof of this concept was obtained with two acid-sensitive doxorubicin prodrugs, i.e. the (4-maleimidophenylacetyl)hydrazone derivative of doxorubicin and the (6-maleimidocaproyl)hydrazone derivative of doxorubicin

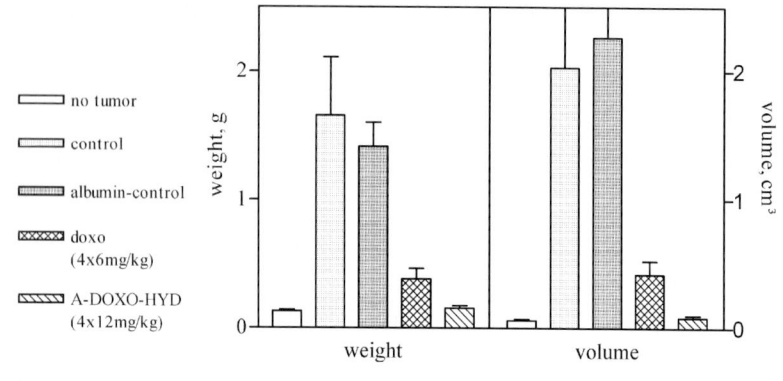

no tumor
control
albumin-control
doxo
(4x6mg/kg)
A-DOXO-HYD
(4x12mg/kg)

A

weight   volume

	Average number of lung metastases
Control	248
Albumin control	408
Doxorubicin (4 x 6 mg/kg)	94
A-DOXO-HYD (4 x 12 mg/kg)	2

B

Control

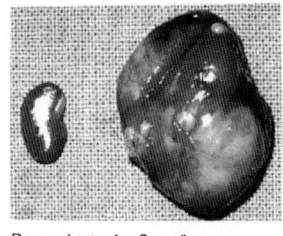
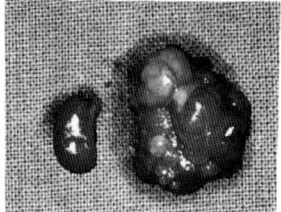

Doxorubicin 4 x 6 mg/kg

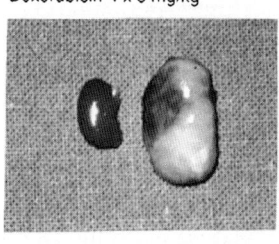

A-DOXO-HYD 4 x 12 mg/kg

C

◄ **Fig. 7** Therapeutic effects of doxorubicin (4×6 mg/kg) and A-DOXO-HYD (4×12 mg/kg) on kidney tumor weight and kidney tumor volume (**A**), and on the number of lung metastases (**B**). **C**: Representative photographic images of control kidney (*right*) and tumor cell treated kidney (*left*) of two mice from the control group, the doxorubicin treated group, and the A-DOXO-HYD treated group. Body weight loss in both treatment groups was comparable (–10 and –12%, respectively)

(DOXO-EMCH) (Fig. 8) which are rapidly and selectively bound to circulating albumin within a few minutes. They are distinctly superior to the parent compound doxorubicin in murine tumor models [29, 30].

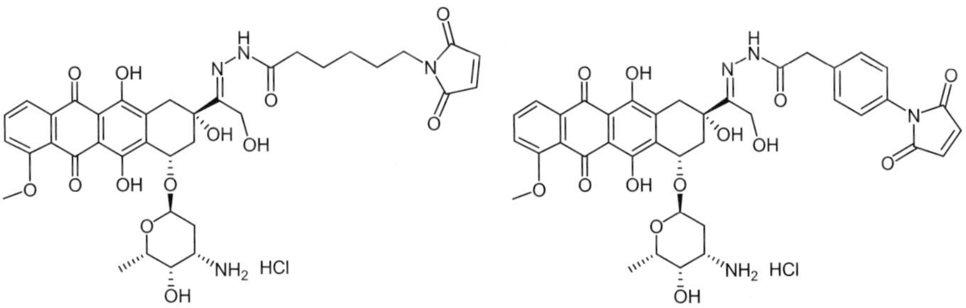

Fig. 8 Structure of (6-maleimidocaproyl)hydrazone (*left*) and (4-maleimidophenylacetyl) hydrazone derivative of doxorubicin (*right*, investigated in a prodrug concept that exploits endogenous albumin as a drug carrier

In a murine renal cell carcinoma model, both prodrugs were superior to free doxorubicin, the clinical standard, with regard to antitumor efficacy and toxicity. In addition, DOXO-EMCH showed a superior activity in two mamma carcinoma xenograft models in nude mice (MDA-MB 435; MCF-7) at equitoxic dose. Therapy with DOXO-EMCH dramatically improved the efficacy of doxorubicin in all three animal models, exhibiting a maximum tolerated dose, which was approximately four times higher than for free doxorubicin.

DOXO-EMCH was selected as the investigational product for clinical evaluation, after toxicology studies in mice, rats, and dogs had shown that DOXO-EMCH exhibits a 2-fold to 5-fold increase in the maximum tolerated dose (MTD) in these animals, when compared to conventional doxorubicin [31]. A 4-cycle intravenous study of DOXO-EMCH at dose levels of 4×2.5, 4×5.0 or 4×7.5 mg/kg doxorubicin equivalents in rats revealed an approximate 3-fold decrease in side effects on the bone marrow system when compared to 4×2.5 mg/kg doxorubicin. The effects on the testes, thymus, and spleen were comparable between both drugs at equitoxic dose, but with a clear indication for recovery in the animals treated with DOXO-EMCH. A no-observable adverse effect level (NOAEL) for DOXO-EMCH of 4×2.5 mg/kg doxorubicin equivalents was established in this study. This dose is equivalent to the MTD of doxorubicin in rats.

In a six-week long, 2-cycle study in Beagle dogs (intravenous administration of DOXO-EMCH at dose levels of 1.5, 3.0, or 4.5 mg/kg doxorubicin
equivalents) only temporary effects on hematology, urinary function, and
histopathology in mid- and/or high-dose animals were observed. The low
dose of 2×1.5 mg/kg was considered to be the NOAEL in this study, which
is equivalent to twice the MTD of doxorubicin in Beagle dogs [31].

DOXO-EMCH has also shown a significant decrease in chronic cardiotoxicity at equimolar, as well as equitoxic doses, compared to doxorubicin in a rat
model [32]. Details of the phase I study, carried out with this prodrug, are
summarized in Chap. 7.

5.2
Doxorubicin Conjugates with Lactosaminated Albumin

A further application of albumin for drug delivery is liver targeting using albumin conjugates containing galactose residues. These neoglycoprotein conjugates are designed to selectively enter hepatocytes by binding to the asialoglycoprotein receptor, with subsequent internalization and degradation of the
carrier in lysosomes, thus circumventing extrahepatic side effects, such as
neurotoxicity of antiviral nucleoside analogues in the treatment of chronic viral hepatitis. The validity of this therapeutic strategy has been demonstrated
in a clinical study, where adenine arabinoside monophosphate (ara-AMP)
conjugated to lactosaminated albumin exerted an antiviral effect comparable to the free drug, without producing any major side effects, including the
severe neurotoxicity of free ara-AMP [33].

Fiume and co-workers adapted their carrier system for a potential treatment
of hepatocellular carcinoma [34–36]. In a study on needle biopsies of 60 human
hepatocellular carcinoma, the asialoglycoprotein receptor was histochemically
detected in 80% well-differentiated and in 20% poorly differentiated forms
of the tumor [37]. This forms the basis for exploiting the asialoglycoprotein
receptor as a molecular target for the selective delivery of drugs to hepatocellular carcinoma. In line with this rationale, the (6-maleimidocaproyl)hydrazone
derivative of doxorubicin (DOXO-EMCH) was coupled to a thiolated form of
lactosaminated human albumin (L-HSA) (see Fig. 9).

The resulting conjugate, L-HSA-DOXO, achieved a very efficient targeting of
the drug to the liver of treated mice, with doxorubicin concentrations reaching
levels 7–20 times higher than those raised in extrahepatic tissues [34].

In further experiments for treatment of hepatocellular carcinoma induced
in mice by N,N'-diethylnitrosamine, L-HSA-DOXO significantly inhibited tumor growth without decreasing body weights (see Fig. 10) [36] at a dose of
4×1 mg/kg doxorubicin equivalents.

In contrast, free doxorubicin administered at the same dose as the coupled
drug, did not affect tumor growth, and it produced a significant decrease
in the body weight of the treated animals. Experiments in healthy rats have

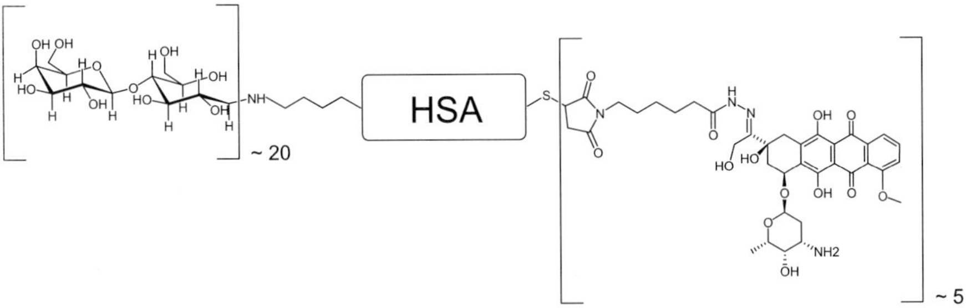

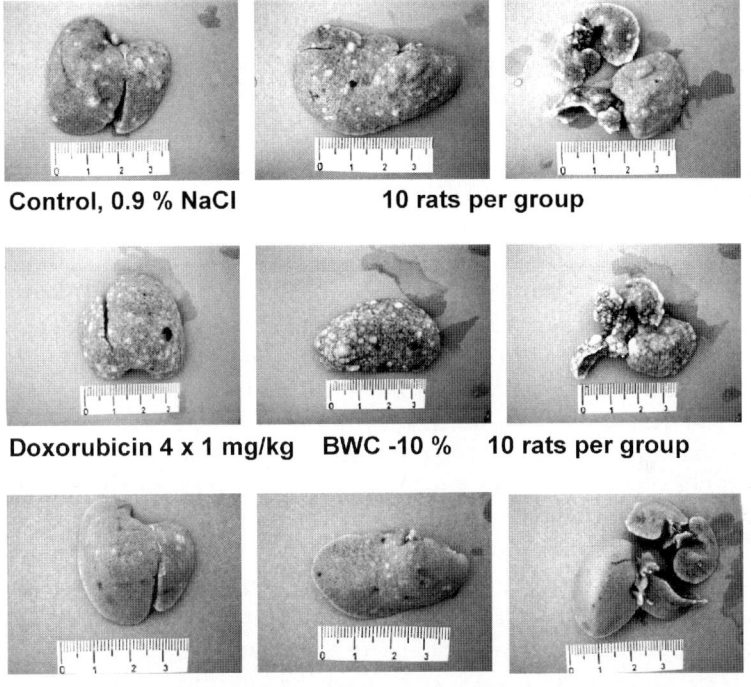

Fig. 9 Structure of L-HSA-DOXO, an albumin conjugate, bearing, on average, 20 molecules of galactose and 5 molecules of DOXO-EMCII, targeting the asialoglycoprotein receptor

Control, 0.9 % NaCl   10 rats per group

Doxorubicin 4 x 1 mg/kg   BWC -10 %   10 rats per group

L-HSA-DOXO 4 x 1 mg/kg   BWC +1%   10 rats per group

Fig. 10 Efficacy of doxorubicin and L-HSA-DOXO in a chemically induced hepatocellular carcinoma model. Representative images of liver tumors for the control group, doxorubicin treated goup, and L-HSA-DOXO treated group

shown that even a dose of 4 × 2 mg/kg L-HSA-DOXO (which is twice the dose of that used in the therapeutic model), produces essentially no liver toxicity, indicating an excellent therapeutic index for the novel conjugate [35].

6
Doxorubicin Prodrugs with Synthetic Polymers

In 1975, Ringsdorf proposed a general schematic design of the drug delivery system for low-molecular weight drugs, using synthetic polymers [38, 39]. One to several drug molecules are bound to a polymeric backbone through a spacer, which incorporates a predetermined breaking point, to ensure release of the drug after cellular uptake of the conjugate. The system can also contain solubilizing groups or targeting moieties, which render the conjugate biorecognizable. Inspired by this pioneering work, a great number of anti-cancer drug-polymer conjugates with different macromolecular carriers have been developed in the last three decades. The vast majority of them are applications consistent with the Ringsdorf's model [39].

This chapter will focus on doxorubicin conjugated with poly(ethylene glycol) (PEG), copolymers of N-(2-hydroxypropyl)methacrylamide (HPMA,) and dendritic polymers, since these polymers have primarily been used for obtaining acid-sensitive prodrugs with doxorubicin.

6.1
Doxorubicin Conjugates with Poly(Ethylene Glycol)

Pendri and colleagues [40] introduced doxorubicin-PEG conjugates with non-cleavable amino bonds or hydrazide bonds, as well as a derivative with an acid-sensitive hydrazone bond. The prodrugs were prepared by reductive amination of PEG (MW 5 kDa) aldehyde with doxorubicin, or by reaction of PEG carbazate with doxorubicin, and, in the case of the hydrazide derivative, by subsequent reduction of the formed hydrazone. In vitro studies against two murine cell lines showed that the prodrug with the cleavable hydrazone bond reached IC_{50} values that were slightly higher, but still in the range of the parent drug.

While early studies focused on PEG-conjugates with low-molecular weight polymer backbones, recent approaches were carried out using PEG of high-molecular weight, considering that the renal clearance of these polymers occurs at much slower rates, leading to prolonged half-lives of the conjugates [41]. In 1999, Rodrigues et al. published various PEG conjugates of doxorubicin incorporating acid-sensitive carboxylic hydrazone bonds, which were designed to be cleaved in the acidic environment of lysosomal or endosomal compartments after cellular uptake (see Fig. 11) [42].

The prodrugs were similar in their design to those developed by Pendri and colleagues [40]. In contrast to former studies, Rodrigues employed high-molecular weight (m)PEGs (MW 20 and 70 kDa), and varied the chemical nature of the carboxylic hydrazone bond [aromatic benzoyl (Hyd_1) to aliphatic phenylacetyl (Hyd_2) spacer]. Incubation studies revealed that at pH 7.4 all prodrugs were stable, whereas at pH 5 the conjugates re-

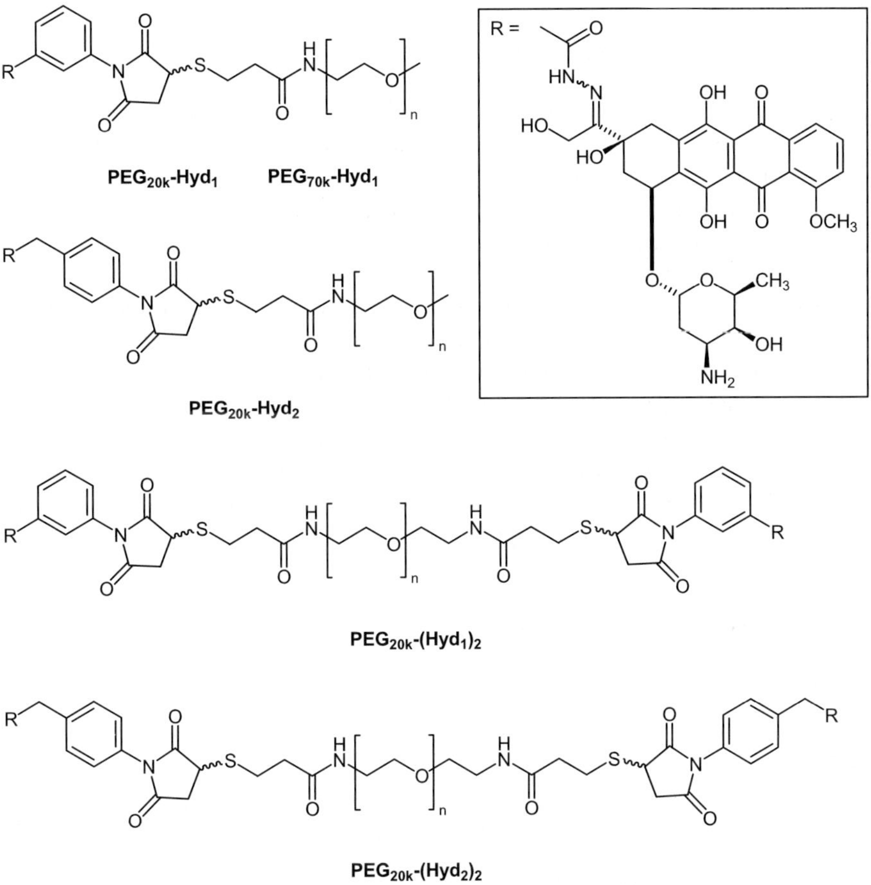

Fig. 11 Structure of various acid-sensitive PEG conjugates of doxorubicin

leased free doxorubicin with half-lives ranging from 2–30 h. Remarkably, the rate of hydrolysis depended on the molecular weight of the polymer backbone with the high-molecular weight prodrug PEG_{70k}-Hyd_1 being significantly more stable to acidic conditions than the PEG_{20k} analogue. Furthermore, it was observed that the spacer group influenced the cleavage rate. Hydrazones with the aromatic benzoyl spacer exhibited a pronounced lability over the phenylacetyl congener. In vitro studies showed that all PEG conjugates with carboxylic hydrazone bonds were cytotoxic against three human cell lines. Although IC_{70} values were 1–2 order of magnitudes above those obtained with the parent drug, there was a correlation between acid-lability and cytoxicity. The PEG_{70k} conjugate was up to 20-fold less active than its PEG_{20k} analogue, and conjugates with the benzoyl spacer proved to be more cytotoxic than those with the phenylacetyl group.

6.2
Doxorubicin Conjugates with HPMA Copolymers

Water-soluble synthetic polymers based on N-(2-hydroxypropyl)methacryl-amide (HPMA), first developed by Kopecek et al. as a plasma expander [43], are non-toxic, non-immunogenic, and non-biodegradable macromolecules. In the early 1980s, the research groups of Kopecek and Duncan recognized the potential and the advantages of HPMA copolymers as macromolecular carriers for anti-cancer drugs such as doxorubicin.

Non-targeted polymeric doxorubicin, in which the drug is bound through either acid-cleavable carboxylic hydrazone bonds or *cis*-aconityl spacers, was reported by Ulbrich and Rihova et al. [44–50]. Several HPMA copolymer conjugates of the drug were synthesized (see Fig. 12), varying in their molecular weight as well as the linkers between doxorubicin and the polymer.

Three of them were designed to be enzymatically *and* acid-cleavable (those with the GFLG spacers incorporated). Incubation studies at different pH values revealed that all conjugates were relatively stable in a buffer solution at pH 7.4 [45, 47]. At pH 5.0, the prodrugs with incorporated hydrazone

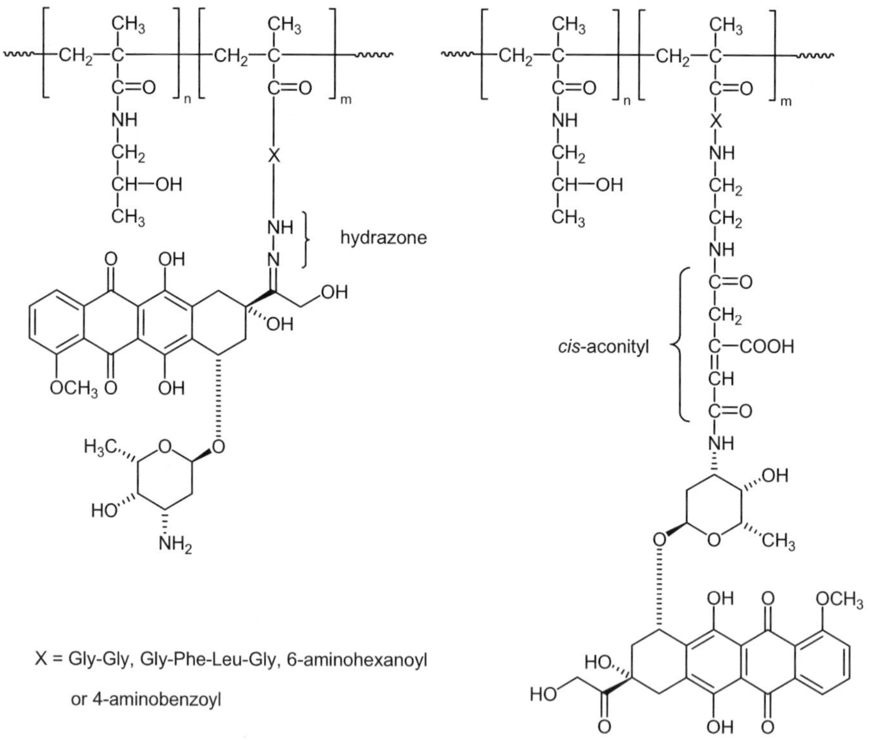

X = Gly-Gly, Gly-Phe-Leu-Gly, 6-aminohexanoyl

or 4-aminobenzoyl

Fig. 12 Structures of acid-sensitive HPMA copolymer conjugates of doxorubicin

bonds released doxorubicin with half-lives of approximately 5 h, whereas the doxorubicin bound via *cis*-aconityl spacers was released at a much slower rate (half-lives > 50 h) [47]. The in vivo efficacy of four of the conjugates with hydrazone linkers was evaluated in mice bearing EL4 T cell lymphoma xenograft models [45, 48]. In this model, the acid-labile doxorubicin polymer conjugate proved to be significantly more active than free doxorubicin as well as PK1, the HPMA-doxorubicin conjugate, in which doxorubicin is bound to the polymer through the GFLG spacer, which is cleaved in lysosomes releasing doxorubicin [45]. Furthermore, in vitro studies against various cell lines [46] revealed that the cytotoxic effect of the acid-cleavable prodrugs is up to two times higher when compared to PK1, and in some cell lines even comparable to the free drug. In the erythroblastoid cell line K 562, with a limited content of lysosomes, acid-labile polymeric doxorubicin exerted pronounced antiproliferative activity.

6.3
Doxorubicin Conjugates with Dendritic Polymers

In the last 10 years, several research attempts have focused on the development of dendrimers or dendritic polymers as high loading capacity carriers for anti-cancer drugs [51]. Dendrimers or dendritic polymers are branched molecules with a defined number of functional groups on their surface, which can be derivatized with anti-cancer agents [52].

An example of a high-loaded PEG conjugate of doxorubicin was published by Fréchet and colleagues [53]. The architecture of the prodrug is based on a star-shaped PEG scaffold (MW 23 kDa), which is grafted with polyester dendrons (see Fig. 13).

Each of the terminal dendrimers bears four molecules of doxorubicin bound through acid-labile hydrazone bonds. Although theoretically up to

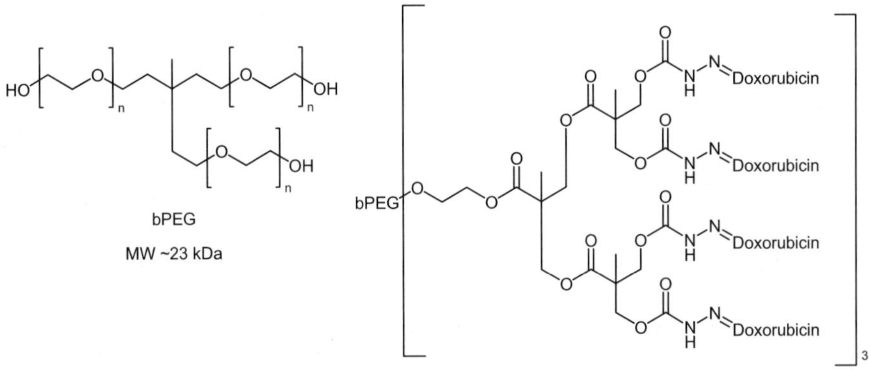

Fig. 13 Structure of an acid-sensitive doxorubicin prodrug containing a PEG scaffold grafted with polyester dendrons

twelve molecules of the drug can be loaded onto the carrier, loading ratios obtained experimentally were much lower, at approximately 50%. The molecular mass of the conjugate, as determined by MALDI-TOF MS, was reported to be 27 kDa. HPLC incubation studies performed with the prodrug, showed a clear correlation between the pH and the rate of drug release [54]. Furthermore, in vitro cytotoxicity was determined using three different cell lines (B16F10, MDA-MB 231, MDA-MB 435). In these experiments, the conjugate was 6-fold to 50-fold less potent than the parent compound.

Problems associated with the use of perfect (monodisperse) dendrimers are obviously related to the synthetic difficulties of achieving sufficiently high molecular masses for passive tumor targeting. Furthermore, attaching drugs to the dendimer peripherally can lead to unpredictable aggregation [55].

Recent research efforts to combine the advantages of linear poly(ethylene glycol) (PEG) and dendritic structures resulted in the development of interesting hybrid materials of different architectures, such as dendronized linear polymers [56], star-like PEG with terminal dendrons [53], or so-called "bow-tie" hybrids [57, 58]. The bow-tie dendrimers synthesized by Gillies et al. consist of two covalently attached polyester dendrons, each of them bearing different terminal groups. One dendron is usually grafted with solubilizing PEG. The other can be loaded with drugs. By using PEG chains of various lengths, it was possible to synthesize well-defined PEG-dendrimer hybrids with low polydispersity and different molecular weights [57]. Due to their lack of toxicity and an advantageous biodistributive profile, selected polymers were considered to be suitable carriers for anti-cancer drugs [59]. Recently, the synthesis of a bow-tie dendrimer conjugated with doxorubicin was reported [60]. The prodrug is based on a pegylated dendritic scaffold (45 kDa) and contains eight PEG chains (5 kDa), and up to sixteen molecules of the drug (8–10% w/w), the latter attached to the dendritic core via acid-sensitive carboxylic hydrazone bonds (Fig. 14).

In an in vivo experiment with C-26 tumor bearing mice, a single i.v. application of the prodrug (20 mg/kg doxorubicin equivalents) produced a complete tumor regression with a 60-day survival of 100%, whereas no cures were achieved with free doxorubicin close to its MTD (6 mg/kg). To the best of my knowledge, this is the first convincing proof that an acid-sensitive dendritic conjugate with an anti-cancer agent holds superior efficacy and tolerability over the parent compound.

6.4
Doxorubicin Conjugates with Other Polymers

Kataoka et al. reported a pH-sensitive supramolecular nanocarrier for doxorubicin, based on biocompatible block-copolymer micelles [61]. Doxorubicin was coupled to a PEO-b-PAsp-copolymer (Fig. 15) through an acid-labile hydrazone linker.

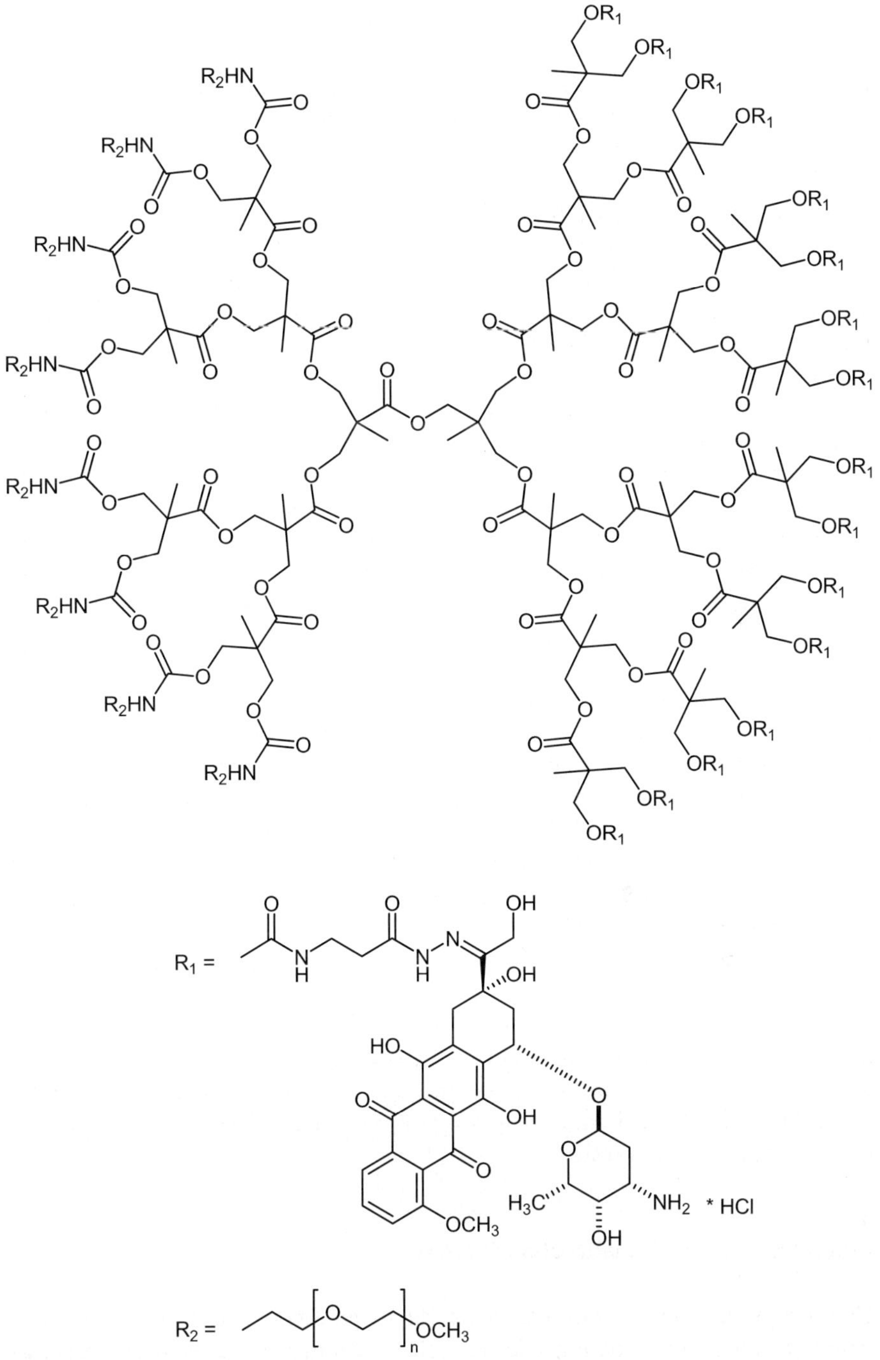

Fig. 14 Structure of a pegylated bow-tie dendrimer loaded with doxorubicin through an acid-sensitive hydrazone bond

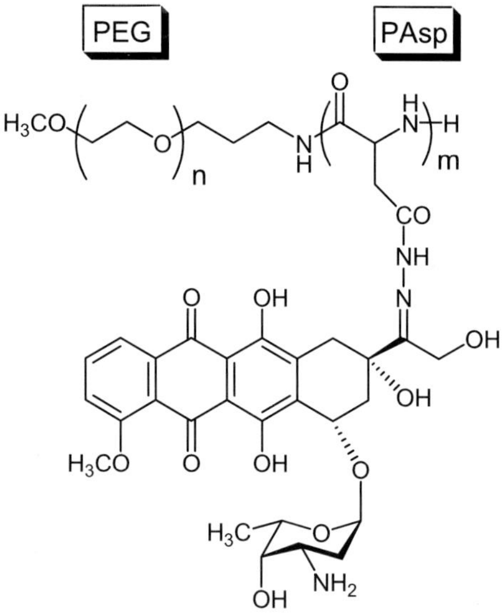

Fig. 15 Doxorubicin block-copolymer conjugate, which self-assembles to form block-copolymer micelles in water. The acid-labile hydrazone bond is cleaved at pH <6, and doxorubicin is released

After spontaneous self-assembly of the drug-loaded supramolecular nanocarrier, the release kinetics clearly demonstrated the effective cleavage of the hydrazone bonds at pH ≤5, with concomitant release of doxorubicin. Under physiological conditions of the cell culture medium (pH ~7), the release of doxorubicin was negligible.

The doxorubicin nanocarrier demonstrated in vitro cytotoxicity against a human small-cell lung cancer cell line (SBC-3) in a time-dependent manner, suggesting cellular uptake via endocytosis. The recent research shows that the pH-sensitive polymeric micelle drug carrier improves antitumor efficacy in tumor bearing mice (C 26 colon adenocarcinoma) at doses of 20 and 40 mg/kg doxorubicin equivalents, compared to 10 mg/kg doxorubicin [62]. At a dose of 10 mg/kg doxorubicin, biodistribution studies showed a 4.2-fold increase in the AUC in the tumor for the doxorubicin micelle over free doxorubicin.

7
Prodrugs of Doxorubicin in Clinical Trials

One acid-sensitive prodrug of doxorubicin has been and one is currently being evaluated in clinical trials. In both cases the prodrug (6-maleimido-caproyl)hydrazone derivative of doxorubicin (DOXO-EMCH) is bound either to the monoclonal antibody BR96 or to endogenous albumin (see Fig. 16).

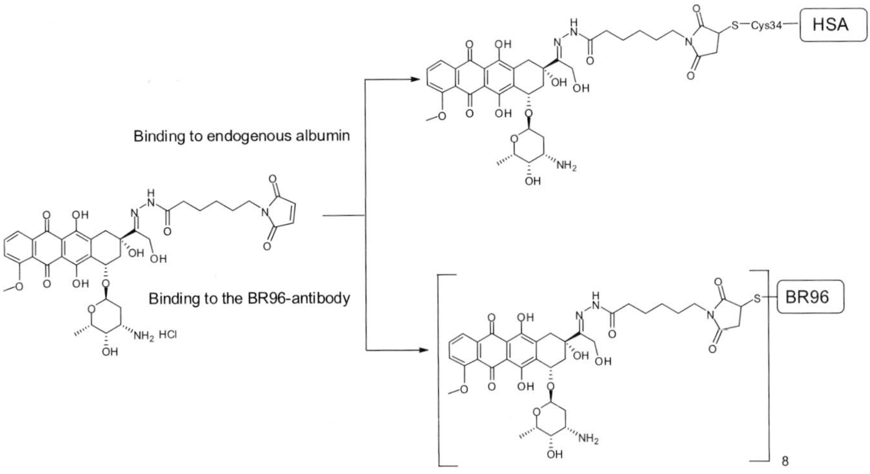

Fig. 16 The (6-maleimidocaproyl)hydrazone derivative of doxorubicin (DOXO-EMCH) bound either to the monoclonal antibody BR96 or to endogenous albumin have been evaluated in clinical trials

In the BR96-doxorubicin conjugate, eight molecules of DOXO-EMCH are bound to the chimeric human/mouse monoclonal antibody BR96, which is specific for Lewis-Y, an antigen that is abundantly expressed on the surface of several human carcinomas, especially in breast cancer [15]. Alternatively, in the passive targeting approach, DOXO-EMCH is bound in situ to the cysteine-34 position of circulating albumin after intravenous application [30, 63]. The results of the clinical trials with these prodrugs are summarized below.

Phase I and Phase II Studies with the BR96-Doxorubicin Immunoconjugate, an Acid-Sensitive Antibody Conjugate of Doxorubicin

The BR96-doxorubicin immunoconjugate (BR96-DOX,) which reacts specifically with the tumor-associated antigen, Lewis-Y, has been evaluated in phase I and phase II studies [18–20]. In the phase I study, the immunoconjugate was administered to 66 patients as an intravenous infusion every 21 days [18]. Doses of BR96-DOX ranged from 66 to 875 mg/m^2, which is equivalent to 2 to 25 mg/m^2 of free doxorubicin. BR96-DOX showed dose-limiting gastrointestinal toxicity at the highest dose. A dose of 700 mg/m^2 (equivalent to 20 mg/m^2 doxorubicin) was recommended for phase II studies. Two patients exhibited a partial response, one with breast carcinoma and the other with gastric carcinoma.

Three phase II studies have been completed with BR96-DOX. A randomized phase II study was performed to evaluate the activity of BR96-DOX in metastatic breast cancer in patients with confirmed sensitivity to the single-agent doxorubicin [20]. Patients with immunohistochemical evidence of Lewis-Y antigen received either $700 \, \text{mg/m}^2$ of BR96-DOX (equivalent to $20 \, \text{mg/m}^2$ DOX) or $60 \, \text{mg/m}^2$ doxorubicin every three weeks. Out of 14 patients receiving BR96-DOX, there was one partial response. However out of 9 patients treated with doxorubicin alone, there were one complete and three partial responses. The cross-reactivity of BR96-DOX with normal gastrointestinal tissue led to prominent toxicities, and probably impaired the delivery of the immunoconjugate to the tumor sites. The low clinical response rate observed in these studies suggests that the dose which could be safely administered every three weeks was insufficient for maintenance of the intratumoral concentration of doxorubicin required to achieve tumor regression.

In the second phase II study, 15 patients with advanced gastric adenocarcinoma expressing the Lewis-Y antigen were treated with $700 \, \text{mg/m}^2$ BR96-DOX every three weeks. Although stable disease was seen in five patients, no objective responses were achieved. Reversible gastrointestinal toxicity, primarily nausea, and emesis were the predominant toxicities.

Finally, a phase II study of BR96-DOX in combination with Taxotere® was initiated, for the treatment of patients with non-small cell cancer, in light of the fact that BR96-DOX demonstrated synergistic effects with paclitaxel and docetaxel in preclinical tumor models (www.seattlegenetics.com) [17]. The data indicate that the combination was well-tolerated, and resulted in improved overall survival, compared to Taxotere® alone, in some patients (www.seattlegenetics.com).

In summary, the shortcoming of this agent was the cross-reactivity of BR96-DOX with normal gastrointestinal tissue, which led to prominent toxicities, and it likely impaired the delivery of the immunoconjugate to the tumor sites. Further development of the antibody conjugate of this drug has been discontinued (Seattle Genetics news release, 6 July 2005).

Phase I Study with DOXO-EMCH, an Albumin-Binding Doxorubicin Prodrug

DOXO-EMCH emerged as a clinical candidate, due to its superior efficacy in several murine tumor models. A 2-fold to 5-fold increase in the maximum tolerated dose and a low cardiotoxic potential were observed, when compared to doxorubicin [30–32, 63, 64]. DOXO-EMCH is selectively bound to the cysteine-34 position of endogenous albumin within a few minutes after intravenous administration. It contains an acid-sensitive hydrazone linker, which allows doxorubicin to be released either extracellularly, in the slightly acidic environment often present in tumor tissue, or intracellularly, in acidic endosomal or lysosomal compartments after cellular uptake of the conjugate by the tumor cell. In the phase I study, a starting dose of $20 \, \text{mg/m}^2$ doxorubicin equivalents was chosen, and 41 patients with advanced cancer were treated at the dose levels of $20–340 \, \text{mg/m}^2$ doxorubicin equivalents [64]. Treatment with DOXO-EMCH was well-tolerated up to $200 \, \text{mg/m}^2$, without manifestation of drug-related side effects. Myelosuppression (grade 1–2) and mucositis (grade 1–2) were the predominant adverse effects at dose levels of 260 mg/m^2. Myelosuppression (grade 1–3) and mucositis (grade 1–3) became dose-limiting adverse effects, at 340 mg/m^2. The expected spectrum of toxicities known from doxorubicin was observed as DLT: Neutropenia, neutropenic fever, and mucositis/stomatitis. No clinical signs of congestive heart failure were observed. In terms of pharmokinetics, the albumin-bound form of DOXO-EMCH has a large AUC, a small volume of distribution and low clearance, compared to doxorubicin [64].

Thirty of 41 patients were assessed for response analysis. Three patients (10%) showed partial response. Fifteen patients (50%) showed stable disease at different dose levels ranging from 2 to 15 weeks. Twelve patients (40%) had evidence of tumor progression. 3 anthracycline-naive patients had partial response lasting for 80, 24 and 17 weeks, respectively. A patient with small-cell lung cancer, pretreated with etoposide and cisplatin, who received six courses of 180 mg/m^2 DOXO-EMCH, achieved an excellent tumor control with time to progression of 18 months. A patient with liposarcoma, pretreated with an angiogenesis inhibitor, was treated with 260 mg/m^2 DOXO-EMCH, and reached a partial remission with time to progression of 17 weeks. A patient with metastatic breast cancer, pretreated with adjuvant CMF, different hormonal treatments was treated with 340 mg/m^2 DOXO-EMCH, and reached a partial remission, and time to progression of 24 weeks.

The recommended dose for phase II studies of DOXO-EMCH is 200–260 mg/m^2 doxorubicin equivalents, which is a 3-fold to 4.5-fold increase over a standard dose of 60 mg/m^2 free doxorubicin. A phase II study of small cell lung cancer (SCLC) has been initiated in June 2007 to assess the antitumor potential of DOXO-EMCH, which has been renamed INNO-206 (see www.innovivepharmaceuticals.com).

In summary, both prodrugs have been investigated preclinically and clinically. A shift in the maximum tolerated dose for both prodrugs over free doxorubicin was noted in preclinical mice models. For DOXO-EMCH (i.v.) [30] this represents an approximate 4.5-fold increase, and for the BR96-doxorubicin conjugate a 2.5-fold increase. In addition, both formulations were superior to doxorubicin, and were able to induce complete remissions in the tumor models studied [15, 30]. The doses needed to achieve complete remissions were higher for DOXO-EMCH than for the BR96-doxorubicin immunoconjugate.

In clinical trials, the shift in the MTD for DOXO-EMCH correlated with that observed preclinically (260 mg/m^2 doxorubicin equivalents for DOXO-EMCH administered as the MTD to humans, compared to standard dose of doxorubicin of 60 mg/m^2), but not for the BR96-doxorubicin immunoconjugate, where the MTD was already reached at 15 mg/m^2 doxorubicin equivalents, due to severe gastrointestinal toxicity, likely caused by the cross-reactivity with the respective normal tissue expressing the target antigen.

References

1. Haag R, Kratz F (2006) Angew Chem Int Ed Engl 45:1198
2. Pimm MV (1988) Crit Rev Ther Drug Carrier Syst 5:189
3. Kratz F, Beyer U (1998) Drug Delivery 5:281
4. Rooseboom M, Commandeur JN, Vermeulen NP (2004) Pharmacol Rev 56:53
5. Gianni L, Grasselli G, Cresta S, Locatelli A, Vigano L, Minotti G (2003) Cancer Chemother Biol Response Modif 21:29
6. Minotti G, Menna P, Salvatorelli E, Cairo G, Gianni L (2004) Pharmacol Rev 56:185
7. Kratz F, Warnecke A, Schmid B, Chung DE, Gitzel M (2006) Curr Med Chem – Anti-Cancer Agents 13:477
8. Jain RK (1987) Cancer Metastasis Rev 6:559
9. Jain RK (1999) Annu Rev Biomed Eng 1:241
10. Mukherjee S, Ghosh RN, Maxfield FR (1997) Physiol Rev 77:759
11. Tannock IF, Rotin D (1989) Cancer Res 49:4373
12. Willner D, Trail PA, Hofstead SJ, King HD, Lasch SJ, Braslawsky GR, Greenfield RS, Kaneko T, Firestone RA (1993) Bioconjugate Chem 4:521
13. Kaneko T, Willner D, Monkovic I, Knipe JO, Braslawsky GR, Greenfield RS, Vyas DM (1991) Bioconjug Chem 2:133
14. Trail PA, Willner D, Lasch SJ, Henderson AJ, Greenfield RS, King D, Zoeckler ME, Braslawsky GR (1992) Cancer Res 52:5693
15. Trail PA, Willner D, Lasch SJ, Henderson AJ, Hofstead S, Casazza AM, Firestone RA, Hellstrom I, Hellstrom KE (1993) Science 261:212
16. Trail PA, Willner D, Knipe J, Henderson AJ, Lasch SJ, Zoeckler ME, TrailSmith MD, Doyle TW, King HD, Casazza AM, Braslawsky GR, Brown J, Hofstead SJ, Greenfield RS, Firestone RA, Mosure K, Kadow KF, Yang MB, Hellstrom KE, Hellstrom I (1997) Cancer Res 57:100
17. Trail PA, Willner D, Bianchi AB, Henderson AJ, TrailSmith MD, Girit E, Lasch S, Hellstrom I, Hellstrom KE (1999) Clin Cancer Res 5:3632
18. Saleh MN, Sugarman S, Murray J, Ostroff JB, Healey D, Jones D, Daniel CR, Lebherz D, Brewer H, Onetto N, LoBuglio AF (2000) J Clin Oncology 18:2282
19. Ajani JA, Kelsen DP, Haller D, Hargraves K, Healey D (2000) Cancer J 6:78
20. Tolcher AW, Sugarman S, Gelmon KA, Cohen R, Saleh M, Isaacs C, Young L, Healey D, Onetto N, Slichenmyer W (1999) J Clin Oncol 17:478
21. King HD, Yurgaitis D, Willner D, Firestone RA, Yang MB, Lasch SJ, Hellstroem KE, Trail PA (1999) Bioconjugate Chem 10:279
22. King HD, Dubowchik GM, Mastalerz H, Willner D, Hofstead SJ, Firestone RA, Lasch SJ, Trail PA (2002) J Med Chem 45:4336
23. Kratz F, Beyer U, Collery P, Lechenault F, Cazabat A, Schumacher P, Falken U, Unger C (1998) Biol Pharm Bull 21:56
24. Kratz F, Beyer U, Roth T, Tarasova N, Collery P, Lechenault F, Cazabat A, Schumacher P, Unger C, Falken U (1998) J Pharm Sci 87:338
25. Kratz F, Roth T, Fichtner I, Schumacher P, Fiebig HH, Unger C (2000) J Drug Target 8:305
26. Beyer U, Rothern-Rutishauser B, Unger C, Wunderli-Allenspach H, Kratz F (2001) Pharm Res 18:29
27. Drevs J, Hofmann I, Marmé D, Unger C, Kratz F (1999) Drug Delivery 6:89
28. Drevs J, Esser N, Richly H, Skorzec M, Scheulen ME, Unger C, Kratz F (2000) Clin Cancer Res 6:120 (Suppl.)
29. Kratz F, Mueller-Driver R, Hofmann I, Drevs J, Unger C (2000) J Med Chem 43:1253

30. Kratz F, Warnecke A, Scheuermann K, Stockmar C, Schwab J, Lazar P, Drückes P, Esser N, Drevs J, Rognan D, Bissantz C, Hinderling C, Folkers G, Fichtner I, Unger C (2002) J Med Chem 45:5523
31. Kratz F, Ehling G, Kauffmann HM, Unger C (2007) Hum Exp Toxicol 26:19
32. Lebrecht D, Geist A, Ketelsen UP, Haberstroh J, Setzer B, Kratz F, Walker UA (2007) Int J Cancer 120:927
33. Fiume L, Di Stefano G, Busi C, Mattioli A, Bonino F, Torrani-Cerenzia M, Verme G, Rapicetta M, Bertini M, Gervasi GB (1997) J Viral Hepat 4:363
34. Di Stefano G, Kratz F, Lanza M, Fiume L (2003) Digestive Liver Disease 35:428
35. Di Stefano G, Derenzini M, Kratz F, Lanza M, Fiume L (2004) Liver Int 24:246
36. Fiume L, Bolondi L, Busi C, Chieco P, Kratz F, Lanza M, Mattioli A, Di Stefano G (2005) J Hepatol 43:645
37. Trerè D, Fiume L, Badiali De Giorgi L, Di Stefano G, Migaldi M, Derenzini M (1999) Br J Cancer 81:404
38. Ringsdorf H (1975) J Polym Sci, Polym Symp 51:135
39. Gros L, Ringsdorf H, Schupp H (1981) Angew Chem 93:311
40. Pendri A, Gilbert CW, Soundararajan S, Bolikal D, Shorr RGL, Greenwald RB (1996) J Bioact Compat Polym 11:122
41. Murakami Y, Tabata Y, Ikada Y (1997) Drug Delivery 4:23
42. Rodrigues PCA, Beyer U, Schumacher P, Roth T, Fiebig HH, Unger C, Messori L, Orioli P, Paper DH, Mülhaupt R, Kratz F (1999) Bioorg Med Chem 7:2517
43. Sprincl L, Exner J, Sterba O, Kopecek J (1976) J Biomed Mater Res 10:953
44. Choi WM, Kopeckova P, Minko T, Kopecek J (1999) J Bioact Compat Polym 24:447
45. Etrych T, Jelinkova M, Rihova B, Ulbrich K (2001) J Controlled Release 73:89
46. Rihova B, Etrych T, Pechar M, Jelinkova M, Stastny M, Hovorka O, Kovar M, Ulbrich K (2001) J Controlled Release 74:225
47. Ulbrich K, Etrych T, Chytil P, Jelinkova M, Rihova B (2003) J Controlled Release 87:33
48. Ulbrich K, Etrych T, Chytil P, Pechar M, Jelinkova M, Rihova B (2004) Int J Pharm 277:63
49. Kovar M, Kovar L, Subr V, Etrych T, Ulbrich K, Mrkvan T, Loucka J, Rihova B (2004) J Controlled Release 99:301
50. Etrych T, Chytil P, Jelinkova M, Rihova B, Ulbrich K (2002) Macromol Biosci 2:43
51. Lee CC, MacKay JA, Frechet JM, Szoka FC (2005) Nat Biotechnol 23:1517
52. Lee CC, MacKay JA, Frechet JM, Szoka FC (2005) Nat Biotechnol 23:1517
53. Ihre HR, Padilla De Jesus OL, Szoka FC, Frechet JMJ (2002) Bioconjugate Chem 13:443
54. Padilla De Jesus OL, Ihre HR, Gagne L, Frechet JMJ, Szoka FC (2002) Bioconjugate Chem 13:453
55. Malik N, Evagorou EG, Duncan R (1999) Anti-Cancer Drugs 10:767
56. Lee CC, Yoshida M, Frechet JMJ, Dy EE, Szoka FC (2005) Bioconjugate Chem 16:535
57. Gillies ER, Frechet JMJ (2002) J Am Chem Soc 124:14137
58. Gillies ER, Frechet JM (2004) J Org Chem 69:46
59. Gillies ER, Dy E, Frechet JM, Szoka FC (2005) Mol Pharm 2:129
60. Lee CC, Gillies ER, Fox ME, Guillaudeu SJ, Frechet JM, Dy EE, Szoka FC (2006) Proc Natl Acad Sci USA 103:16649
61. Bae Y, Fukushima S, Harada A, Kataoka K (2003) Angew Chem Int Ed Engl 42:4640
62. Bae Y, Nishiyama N, Fukushima S, Koyama H, Yasuhiro M, Kataoka K (2005) Bioconjugate Chem 16:122
63. Kratz F (2007) Expert Opin Investig Drugs 16:855
64. Unger C, Häring B, Medinger M, Drevs J, Steinbild S, Kratz F, Mross K (2007) Clin Cancer Res 13:4858

Top Curr Chem (2008) 283: 99–140
DOI 10.1007/128_2007_12
© Springer-Verlag Berlin Heidelberg
Published online: 24 November 2007

Doxorubicin Conjugates for Selective Delivery to Tumors

Jean-Claude Florent · Claude Monneret (✉)

Laboratoire de Pharmacochimie, Institut Curie,
26 rue d'Ulm, 75248 Paris cedex 05, France
claude.monneret@curie.fr

Abstract With the aim of improving the therapeutic utility of doxorubicin, numerous conjugates or prodrugs have been prepared to be selectively activated at the tumor site while releasing the cytotoxic drug.

Among immuno-conjugates representing a widely studied class of doxorubicin derivatives, the clinical development of cBR96-Dox, undoubtedly the most quintessential derivative, was discontinued due to severe secondary effects. More potent cBR-96 analogues and IMMU-110, another doxorubicin immunoconjugate, are still under study.

Antibody-directed prodrug therapy has been designed to overcome some of the problems associated with the treatment of solid tumors. Concerning the anthracycline-based prodrugs, two glucuronide conjugates have reached the preclinical level, HMR 1826 and DOX-GA3. Both conjugates were subsequently evaluated against several human cancer xenografts without preliminary administration of fusion protein. Among the novelty in ADEPT approaches, one of the most relevant was based on the design of multiple spacer systems.

Closely related to ADEPT, new approaches to selectively deliver prodrug-releasing enzymes in tumor cells have been still studied or proposed by means of gene (GDEPT), polymer (PDEPT), bacteria (BDEPT), or exploiting endogenous carbohydrate-lectin binding (LEAPT).

Activation of conjugates by tumor-associated endogenous enzymes such as prostate specific antigen, plasmin, matrix metalloproteinase, and various extra and intracellular peptidases has also been reported, some of these conjugates like L377,202, a PSA substrate, having reached the clinical level. Doxorubicin peptide conjugates were also designed to be activated by endopeptidase legumain, and extracellular thimet oligopeptidase to deliver Leu-Dox, known to be cleaved intracellularly by peptidase.

A third class of conjugates has been designed for receptor-mediated targeted delivery, including folate, somatostatin, bombesin, LHRH receptors or integrin and lectin.

Transportation of doxorubicin with peptide vectors has been simultaneously investigated to overcome the problem of penetration in the brain or the problem of multidrug resistance.

Keywords Doxorubicin · Targeting · Conjugate · Prodrug · Antibody-receptor

Abbreviations

ADEPT	Antibody-directed enzyme prodrug therapy
BDEPT	Bacterial-directed enzyme prodrug therapy
DNR	Daunorubicin
Dox	Doxorubicin
GDEPT	Gene-directed enzyme prodrug therapy
IC_{50}	50% Inhibitory concentration
i.p.	Intraperitoneal
i.v.	Intravenous
LD	Lethal dose
LEAPT	Lectin-directed enzyme-activated prodrug therapy
Leu	Leucyl
LHRH	Luteinizing hormone releasing receptor
PDEPT	Polymer-directed enzyme prodrug therapy
PSA	Prostate specific antigen
SST	Somatostatin
SSTR	Somatostatin receptor

1
Doxorubicin Immuno-Conjugates

Conjugates of monoclonal antibodies (Mabs) with drugs have been investigated for many years as a potential approach to deliver cytotoxic agents to the tumor target more specifically. The efficacy of drug-MAb immunoconjugates in vitro and in vivo have been several times [1–3] reviewed and besides Mylotarg (gemtuzumab ozogamicin) a conjugate of the anti-CD33 antibody with the highly potent cytotoxic drug, calicheamicin, which has been approved by the FDA, some of them are under clinical studies.

The Bristol-Myers Squibb's group (BMS) has synthesized several immuno-conjugates [4] of doxorubicin by joining MAb and Dox via disulfide or thioether bonds to the Mab moiety and hydrazone bonds to the doxorubicin C-13 carbonyl. One such immunoconjugate that utilizes the C-13 hydrazone and involves an anti-Lewis Y Mab, the cBR96-Dox 1 (Fig. 1), is very stable but release Dox following exposure to the acidic pH of endosomes/lysosomes. This conjugate is selectively internalized by a wide variety of human carcinomas expressing an extended form of Lewis Y antigen, and displayed relevant activity on large human tumor xenografts implanted in mice at doses of 5 to 20 mg/kg in doxorubicin with a selectivity in the range of 8–25. Due to its broad and potent activity against multiple tumor models, cBR96-Dox was developed for clinical trials.

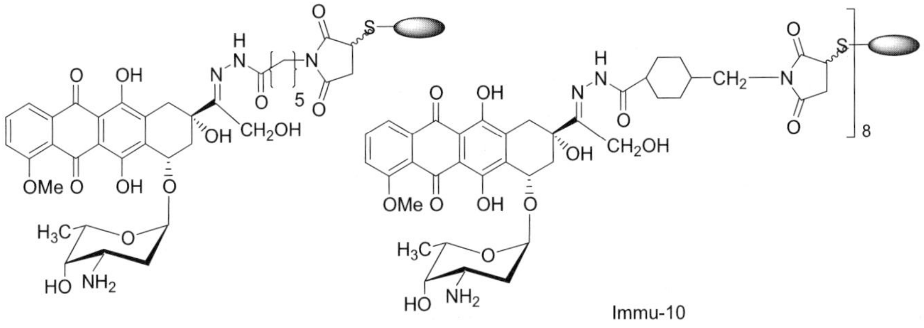

Fig. 1 cBR96-Dox conjugate 1 and immu-110

However, when administered alone in a phase I clinical trial in patients suffering predominantly of metastatic colon and breast cancer and whose tumors expressed the LeY antigen [5], no significant activity was found. Thus, despite tumor stabilization and some partial responses during a randomized phase II study [6] in metastatic breast cancer, secondary effects such as gastrointestinal toxicities, nausea and vomiting with gastritis precluded dose escalation. Although a dramatic increase in regression rates of breast cancers was observed when cBR96-Dox was combined to paclitaxel [7], BMS decided to discontinue further clinical development of cBR96-Dox in breast cancer.

Next this conjugate was licensed to Seattle Genetics [8] and developed under the name of SGN-15. After having been shown that treatment with cBR96-Dox prior to paclitaxel administration resulted in a steady increase in sensitivity to taxanes [9], a randomized, multicenter phase II study of SGN-15 combined with docetaxel including 62 patients with advanced stage of metastatic non-small cell lung cancer was undertaken [10]. Despite the fact that SGN-15 plus docetaxel was well tolerated and active in second and third-line treatment of NSCLC patients, Seattle Genetics decided to discontinue development of SGN-15 on July 2005 [11].

It should be noticed that the BR96-Dox immunoconjugate was also found effective against intracerebral tumors when delivery is enhanced with osmotic disruption of the blood–brain barrier [12] but was not effective against glioma cells.

A series of more potent cBR-96 analogues was synthesized during the 2000–2007 period, involving 5-diacetoxypentyl Dox (DAPDox) 2 and morpholinoDox 3 [13], which are 2–250 times more potent against Dox-sensitive cell lines in vitro, are not substrates for P-glycoprotein and thus do not display MDR phenotype (Fig. 2). Immunoconjugate molar ratios were generally 7.5–8.5 and the linkers released parent drug at lysosomal pH 5 while remaining stable at neutral pH. The BR96-diacetylpentyl Dox displayed an IC_{50} of 0.03 mM against L2987 lung carcinoma cell line (BR96 antigen positive) and was at least 300-fold more potent that a non-binding morpholinoDox control conjugate.

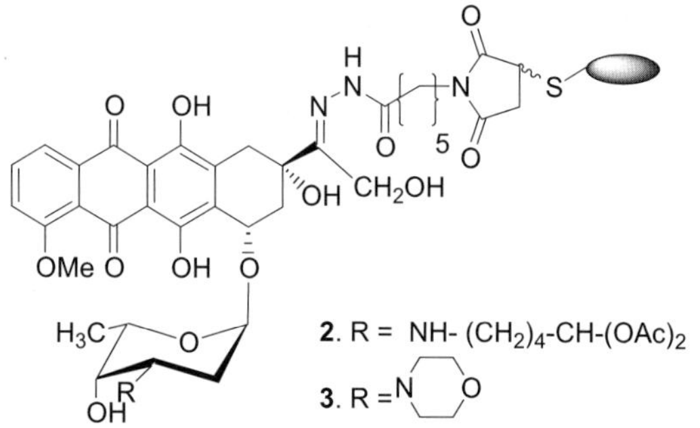

Fig. 2 5-diacetoxypentylDox and morpholineDox

To prepare more potent doxorubicin conjugates, use of a bivalent or branched linker was also designed. Indeed for the usual access to BR96-Dox, the method of conjugation used eight thiol groups, generated by dithiothreitol reduction of four interchain disulfides. This limits molar ratios to a maximum value of 8 mol drug/mol BR96. Immunoconjugates involving branched peptide-doxorubicin linkers as bivalent compound 4 (Fig. 3) increased this molar ratio to 16 mol drug/mol BR96 but [14] as the first immunoconjugates

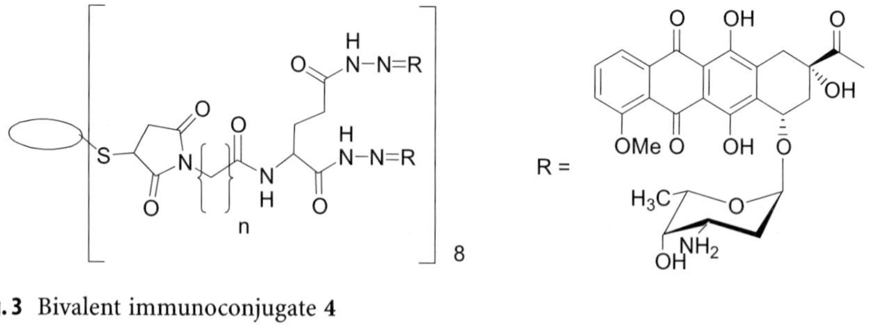

Fig. 3 Bivalent immunoconjugate 4

were prone to noncovalent, dimeric aggregation, a hydrophilic methoxytriethylene glycol chain was subsequently added [15]. These last immunoconjugates were highly potent and immunospecific in vitro. After antigen-specific
internalization, liberation of doxorubicin would occur by hydrolysis of the
methoxytriethylene glycol chain into tumor lysosomes in vitro, and subsequent enzymatic degradation of the peptide linker. Aggregation was well
prevented by the introduction of hydrophilic mTEG residues.

Dox-immunoconjugates containing dipeptide linkages cleavable by lysosomal proteases, such as cathepsin, following internalization of the conjugate,
were also reported. According to Dubowchik et al. [16], drug-dipeptide (Phe-
Lys) linker was still linked to the free thiol groups generated on MAb by
treatment with dithiothreitol (DTT). This bivalent maleimide Dox-containing
(Fig. 4) was next conjugated to chimeric BR96 to give the bivalent conjugate
5 with a drug/MAb mole ratio of 14 : 1 which demonstrated antigen-specific
in vitro cytotoxic effect equipotent to Dox with an IC_{50} of 0.2 µM, against the
antigen-positive L2987 lung carcinoma cell line.

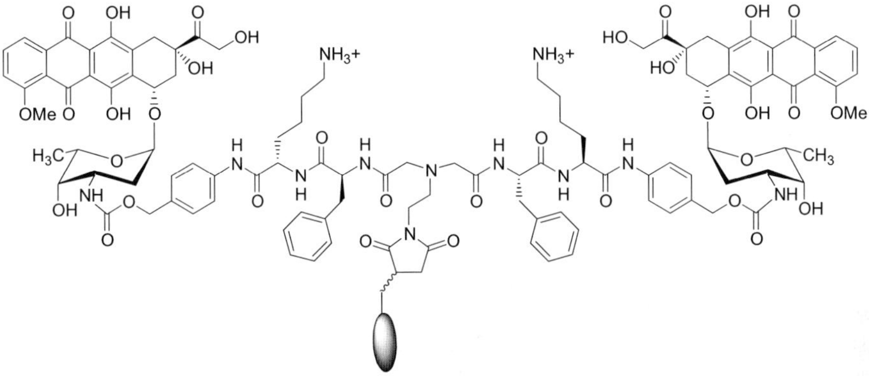

Fig. 4 Bivalent immunoconjugate **5**

On the other hand, dipeptide-based more highly potent conjugates were
prepared (Fig. 5). The purpose was to deliver cyclic derivatives of Dox such
as the morpholino cyanide derivative **6** [17] or the pyrrolidine derivative **8**
arising from cyclization of the *n*-butyl and *n*-pentyl diacetates **7** under esterase hydrolysis [18], both analogues displaying largely higher cytotoxicity
than Dox.

Two approaches (Fig. 6) were followed involving protease-mediated release of **8**. According to the first approach (route A), the drug was released
from the MAb by proteolytic hydrolysis of a peptide linker, prior to the
formation of the reactive pyrrolidine ring. The second approach involved
tethering the molecule through an oxazolidine carbamate which would stable and masked the activating aldehyde group until proteolytic cleavage from
the mAb (route B) occurred. Cleavage of the dipeptide would result in 1,6-

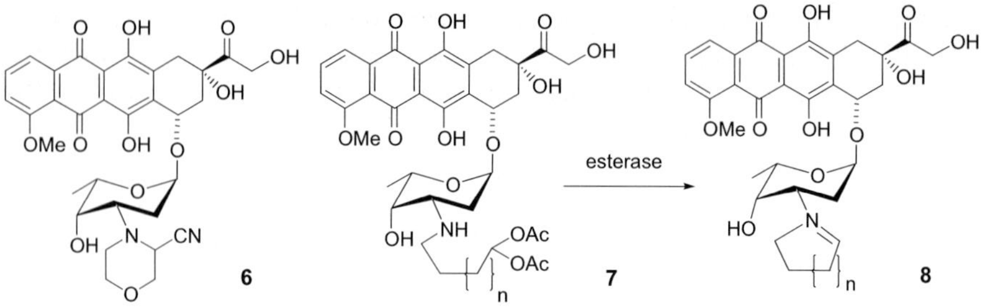

Fig. 5 Doxorubicin analogues, 6, 7 and 8

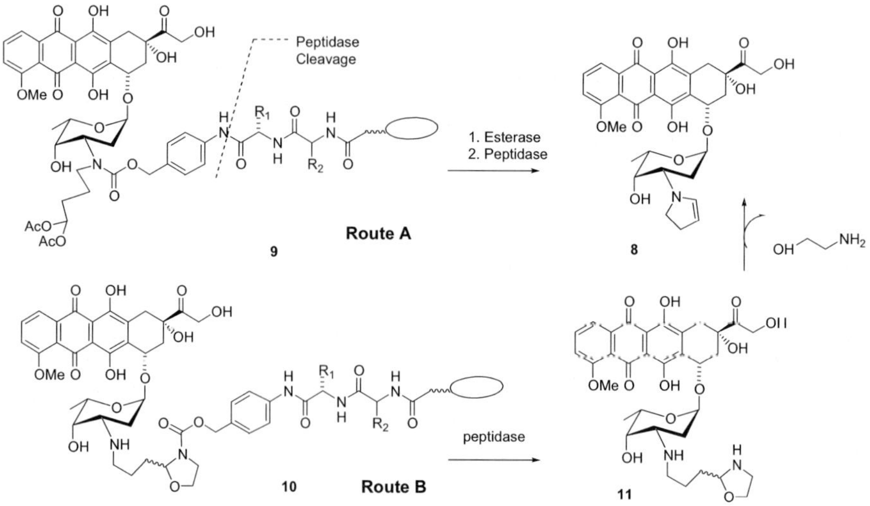

Fig. 6 Mechanism of proteolytic release of doxorubicin

elimination to liberate an intermediate that should rapidly cyclized to afford **8**.

Thus, the two linkers **a** (R_1 = $(CH_2)_3$-NH-$CONH_2$; R_2 = Isopropyl) and **b** (CH_3; R_2 = Isopropyl) were synthesized and conjugated to the chimeric MAbs 1F6 (which binds to CD70 antigen of the TNF receptor superfamily) and cAC10 (which binds to the CD30 antigen), affording **9a,b** and **10a,b**. Both conjugates **9** and **10** were substrates for the lysosomal enzyme cathepsin B and were cleaved at comparable rates. When tested in vitro against CD70$^+$ or CD30$^+$ cell lines, > 40% specificity were observed with values of IC$_{50}$ between 0.4 and 2.3 nM, which means that conjugates were > 70 fold more potent than DOX. On the other hand, the IC$_{50}$ values of conjugates with **4** or **5** as linker were similar suggesting that the active drug was similarly delivered.

IMMU-110, another doxorubicin immunoconjugate (Fig. 1) is structurally closely related to the cBR96-Dox conjugate but involves murine and human versions of the anti-B cell, antibody LL1, targeting CD74, a rapidly internalizing antibody that is highly expressed in several tumors. This conjugate, which carried approximately eight drugs per MAb, has been tested successfully on SCID mice bearing human B-lymphoma xenografts [19]. Moreover, with respect to the high prevalence of CD74 antigen expression in multiple myeloma, IMMU-110 was evaluated [20] in mice and monkeys as well as in a xenograft model of the human multiple myelome cell line (MC/CAR). Treated with a single dose of IMMU-110 as low as 50 μg antibody/mouse (or 1.4 μg of dox/mouse), 5 days post injection of multiple myeloma cells resulted in cure of most mice. No toxicity was observed, including myelotoxicity and cardiotoxicity, up to the maximum single dose tested of 125 mg/kg (3.6 mg/kg of Dox). The therapeutix index of IMMU-110 in the MC/CAR mouse xenograft model was found to be > 50 fold. In nonhuman primates like cynomolgus monkeys, still no acute cardiotoxicity was observed as well as adverse effects to other major organs at doses up to 90 mg/kg, except bone marrow toxicity which appeared at 30 mg/kg.

A MAb-Dox conjugate has been also reported in an attempt to target head and neck cancers and squamous cell carcinoma (SCC). This immunoconjugate directed against hsp47/CBP2 has been prepared [21] by linking the 13-keto position of the drug to the MAb via an hydrazone linker. Biological studies revealed that SCC cells treated with this SPA470-Dox conjugate retained the original binding activity for SCC cells and was significantly more potent that unconjugated Dox, Dox-hydrazone or MAb + Dox. Further demonstrations indicate that SPA470-Dox is effective during hypoxia, condition which influences the expression of hssp47/CBP2 and thus presumes the further utility of the conjugate in treating head and neck cancers.

Hepatocellular carcinoma cell (HepG2) were also targeted by linking doxorubicin to a monoclonal antibody anti-midkine [22] which is a heparin-binding growth factor, preferentially expressed in tumor cells.

Targeting Dox via copolymers linked to MAbs has also been reported. In one case, such a copolymer has been linked to C225 MAb [23] directed against epidermal growth factor receptors (EGFRs) through a polyethylene glycol spacer. Combination chemotherapy and photodynamic therapy has been evaluated in the second case by targeting with OV-TL 16 MAb, directed against ovarian cell lines, a copolymer-bound doxorubicin and mesochlorin [24]. In both cases, enhanced delivery of dox was observed which, besides the role of antibody-targeting, may attributed to enhanced permeability and retention (EPR) effects.

2
Conjugates for Directed-Enzyme Prodrug Therapies

2.1
Antibody-Directed Enzyme Prodrug Therapy (ADEPT)

The antibody-directed prodrug therapy has been designed [25] to overcome some of the problems that were associated with the treatment of solid tumors, including poor penetration of the immunoconjugates, antigen heterogeneity, low drug potency and inefficient drug release. In this two-step therapeutical approach, a mAb-enzyme conjugate is administered in a first phase. Once it localized within the tumor mass and cleared from the systemic circulation, an anticancer prodrug is given in a second phase which is converted to the active drug by the targeted enzyme.

Many enzymes of mammalian and non-mammalian origin have been reported for use within the ADEPT [26–28]. Concerning the drugs, a large panel has been studied, and among them, anthracycline antibiotics. Nevertheless, to our knowledge, concerning the anthracycline-based prodrugs, only two glucuronide conjugates have reached the preclinical level, HMR 1826 and DOX-GA3.

HMR 1826 (Hoechst–Marion–Roussel) is a doxorubicin prodrug (Fig. 7) developed in our laboratory [29] in collaborative work with Dr. Bosslet et al. from Behring Institute. In order to target the enzyme at the tumor cell surface of gastro-intestinal tractus, a fusion protein was elaborated from human β-glucuronidase and a single chain fragment of the humanized mAb anti-CEA [30]. Self-immolative spacer-containing prodrugs were designed on the basis of preliminary results [31].

The main results observed with HMR 1826 were its stability in plasma, its high detoxification with an $IC_{50} = 2\,\mu M$ versus $0.02\,\mu M$ for the doxorubicin and a MTD, which was $> 1200\,mg/kg$ versus $12\,mg/kg$ for DOX. In vivo a 4–12 fold higher doxorubicin concentration in tumors and a five-fold lower drug concentration in normal tissues were measured [32]. Moreover, HMR 1826 was found to be 100-fold less cardiotoxic than doxorubicin [33].

Elevated activity of β-glucuronidase in tumor tissue relative to normal tissue was first observed by Fishman and Anlyan [34] but exploitation has not been widely used except by Connors and Whisson [35], and Double and Workman [36], who reported therapeutic response to aniline mustards in high glucuronidase tumor-bearing mice. Taking into account these data, Bosslet et al. [37] re-examined tumors using enzyme histochemical methods and found that lysosomal β-glucuronidase is liberated in necrotic tumors area, the cells responsible for the liberation of the enzyme being mainly acute and chronic inflammatory cells [38]. Thus, it became evident that glucuronide prodrugs designed for ADEPT could be used in prodrug monotherapy (PMT). This was demonstrated on isolated human lungs perfusion with

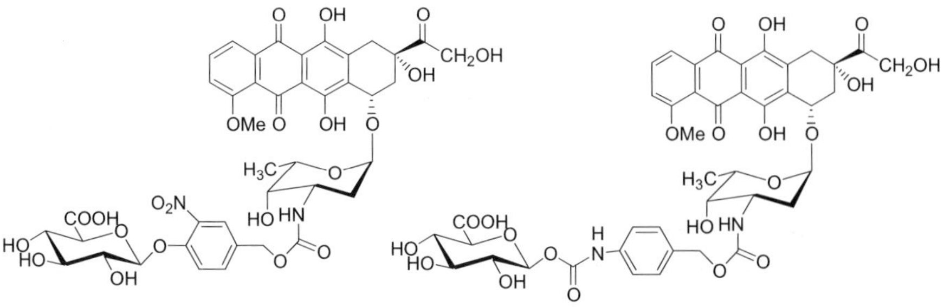

Fig. 7 HMR1826 and DOX-GA3

HMR 1826 [39, 40] by enhanced uptake of doxorubicin into bronchial carcinoma but also on a large panel of human tumor xenografts in nude mice [38].

Interest in HMR 1826 has been underlined by comparison with free Dox and liposomal Dox [41] and by the fact that elevated expression of β-glucuronidase has been observed in pancreatic cancer (100 to 300%) [42, 43]. Nevertheless, development of HMR 1826 was discontinued by decision of Hoescht-Marion-Roussel and then Aventis.

Carbamate-based spacers were also designed and synthesized in the group of Scheeren and evaluated by the group of Haisma et al. [44, 45]. Among the numerous prodrugs they prepared [46], DOX-GA3 was selected for development (Fig. 7).

First experiments [47, 48] were conducted in the presence of an enzyme-conjugate prepared from the murine MAb 323/A3, which is specific for a pan-carcinoma Ep-CAM, and slightly modified human glucuronidase (mGUS). In nude mice bearing s.c. human ovarian cancer xenografts, improved antitumor effects were observed at doses of 500 mg/kg, the maximum tolerated dose of DOX-GA3 being of 500 mg/kg weekly ×2. However, at lower dose of 250 mg/kg, tumor-growth inhibition was not better than that of doxorubicin alone [49]. Next, a fusion protein was prepared [50] consisting of a human single-chain Fv antibody C28 against the epithelial cell adhesion molecule, and the human enzyme β-glucuronidase. The sequences encoding C28 and human β-glucuronidase were joined by a flexible linker. As DOX-GA3 is rapidly excreted by the kidneys, the same group further hypothesized [51] that a slow release of DOX-GA3 from its carboxymethyl ester (DOX-mGA3) by esterase activity in blood would result in improved circulation half-life of DOX-GA3. Not only DOX-mGA3 is synthesized more efficiently than DOX-GA3, but improved pharmacokinetics were observed with this more lipophilic prodrug. It was postulated that this effect may even be more pronounced in man, because of the lower plasma esterase than measured in mice.

Based on the release of glucuronide prodrugs in necrotic area of tumors like HMR 1826, the corresponding daunorubicin-GA3 was successfully eval-

uated against human ovarian cancer xenografts without preliminary administration of fusion protein [52].

In order to improve the efficacy, selectivity and non-immunogeneicity of
ADEPT approaches during the last 10 years, three main directions have been
followed:

- development of non-immunogenic, new fusion protein,
- used of selective enzymes ...,
- design of elongated self-elimination or multiple spacer systems.

Concerning the first direction, β-lactamase is a versatile enzyme that can
activate a variety of anticancer drugs and among them, doxorubicin [53]. The
rationale of using this non-mammalian enzyme was based on the data that
the corresponding prodrugs are stable, non-toxic, and are not substrates for
endogenous human enzymes, avoiding premature activation. However, such
enzymes elicit immune responses in human. Therefore intense efforts have
been made to prepare less immunogenics variant β-lactamase [54] or fusion
proteins such as TAB2.5, which results from the fusion of the MAb CC49 with
the enzyme [55].

β-galactosidase as β-glucuronidase belongs to the class of enzymes of
mammalian origin and thus is expected to be much less immunogenic
than enzymes like lactamase or carboxypeptidase. In the 1990s, prodrugs
of DNR [56–59] as substrates for α or β galactosidase have been synthesized. More potent Dox analogues were also involved in these approaches.
Thus a series of various ω-[bis(acetoxy)]alkyl or ω-(bis(acetoxy)]alkoxyalkyl
derivatives, substituted at the 3'-amino position were designed [60] to be
hydrolyzed in the presence of carboxylate esterases giving N-(5-oxypent-1-
yl)doxorubicin as illustrated (Fig. 8) with the most potent compound of the
series. It was shown that the chain length of the 3'-amino substituents and
the stability of the derived aldehydes to form five- or six-membered carbino-
lamines are critical determinants for the biological activity.

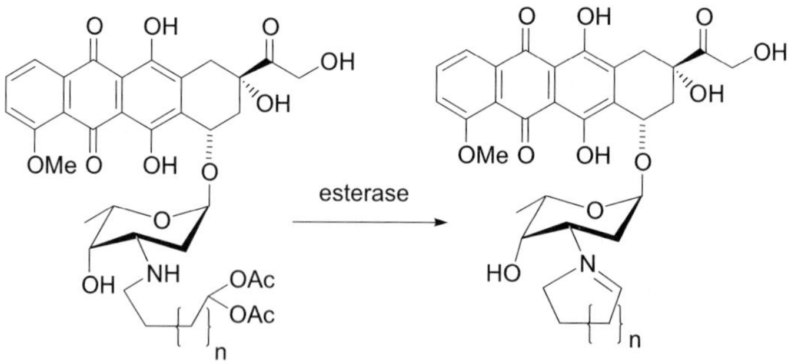

Fig. 8 Bis(acetoxy)alkyl doxorubicin derivatives

More recently, the L49 MAb directed against the p97 antigen on melanomas and carcinomas was chemically conjugated [61] to *E. coli* β-galactosidase in order to activate the daunorubicin conjugate as illustrated in Fig. 9. Activation of this conjugate led to a highly cytotoxic drug with a cytotoxic differential of approximately five orders of magnitude (10 pM versus 0.8 μM), which may be fruitful if very small percentages of conjugate accumulate within the solid tumor.

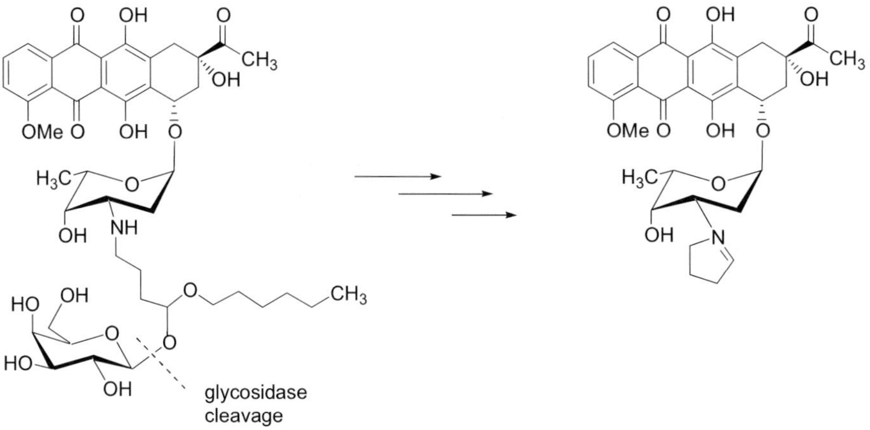

Fig. 9 Hexyloxy-galactosyl-butyl doxorubicin derivative

With regard to more specific enzymes involved in ADEPT, an example with post-proline cleaving endopeptidase was reported. The human prolyl endopeptidase was expressed in *Escherichia coli* and purified. In order to improve its stability and thus to obtain a thermostable enzyme, a single amino-acid mutation (Glu289 → Gly) was carried out, resulting in a half-life of 16 h at 37 °C in phosphate buffer. The purified prolyl endopeptidase mutant was chemically coupled to mAb SIP(L19), which recognizes DE-B domain of fibronectin. On the other hand, N-protected glycine-proline dipeptide was covalently coupled to doxorubicin. However, whereas a melphalan prodrug was efficiently activated, the doxorubicin prodrug remained stable, probably due to steric hindrance [62].

Concerning the third direction, since a successful in vivo selective prodrug depends on an efficient activation by the site-specific enzyme, Shabat et al. designed and synthesized self-immolative dendrimers. The first generations were based on adaptable molecules, 4-hydroxymandelic acid [63] and above all, 2,6-bis(hydroxymethyl)-*p*(-cresol) [64, 65] that have three functional groups, and thus represents a potential platform for a multiple prodrug. In the case of the cresol derivative, the two hydroxymethyl groups (Fig. 10) were attached through a carbamate linkage to the enzyme target and

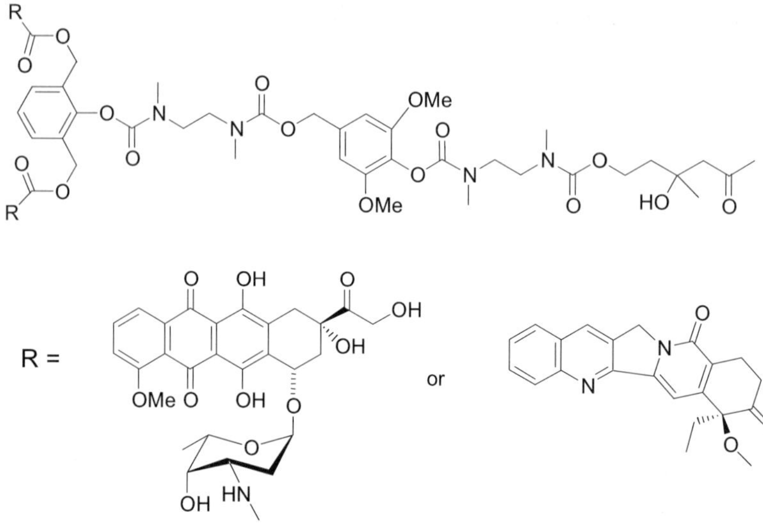

Fig. 10 Bis-hydroxy-*p*-cresol conjugates

the phenol function was linked to the drug through a short spacer, *N,N'*-dimethylethylenediamine.

As proof of the concept, they designed a pilot system for which catalytic antibody 38C2 was selected as the cleaving enzyme [66]. The antibody catalyzes a sequence of retroaldol retro-Michael cleavage reactions as illustrated as follows, affording two doxorubicin units by a single cleavage. Moreover, substrates of the enzyme are not recognized by human enzymes.

A single-triggered trimeric prodrugs system was further designed [67] that proceeds through a triple quinone methide rearrangement under physiological conditions. They synthesized two prodrugs, one containing exclusively three camptothecin units, the other containing three different drug molecules, camptothecin, etoposide and doxorubicin. The same system of delivery as above with retro-aldol and retro-Michael activation by antibody 38C2 was able to simultaneously release the three different chemotherapeutic drugs at the same site. A heterodimeric prodrug combining a camptothecin unit and a doxorubicin unit was also prepared.

It must be noted that de Groot et al. [68] and Szalai et al. [69] have also designed dendritic units that are able to release several drug molecules with a single enzymatic cleavage but no application for doxorubicin release were involved in both papers.

As elongated spacer systems could decrease steric hindrance to a larger extent than conventional spacer, de Groot et al. designed [70] different types of elongated cascade spacer. Three Dox conjugates that contain one, two, or three self-immolating spacers as depicted in Fig. 11 were thus synthesized. All these prodrugs were stable in 0.1 M Thris/HCl buffer (pH 7.3) for 3 days at

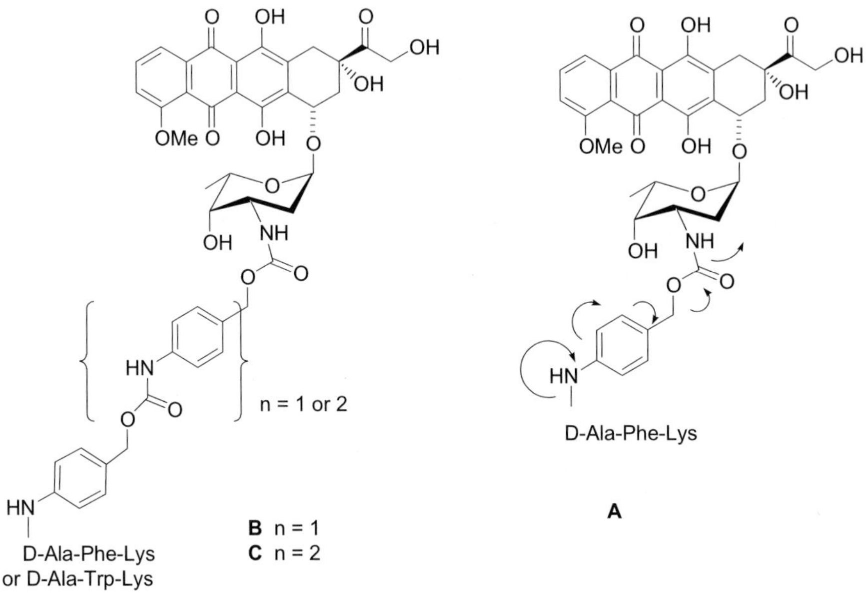

Fig. 11 One, two, or three spacer-containing Dox conjugates

37 °C. In the presence of plasmin, the time of enzymatic hydrolysis to reach a 50% conversion to Dox was in the following order: A > B > C. Thus, the triple spacer prodrug C shows the highest enzymatic activation rate. Comparison to prodrugs containing a 2′-carbonate linkage (no represented) indicates that a 2′-carbamate function as in A, B, and C increase drug stability.

2.2
Gene-Directed Prodrug Therapy (GDEPT)

GDEPT or suicide gene therapy is another approach to target cytotoxic agent to tumor cells in a selective and specific manner [71]. It has been proposed as an alternative to solve the major problem of tumor gene therapy, the low transduction efficiency of the vectors.

It only differs from ADEPT in delivering the enzyme in the tumor cells by introducing a gene which encodes for this enzyme, instead of using a conjugate or a fusion protein. This means that prodrugs must be able to penetrate across tumor cell membranes instead of being localized at the cell surface in ADEPT and therefore, must be lipophilic [72].

However, among the few examples of GDEPT approaches with Dox, two of them were based upon use of a gene encoding for secreted human β-glucuronidase, so that previous hydrophilic prodrugs such as HMR 1826 or DOX-GA3 were involved in assays. The first example was reported by Weyel et al. [73] who demonstrated that tumor cells transduced with a secreted form of β-glucuronidase (s-β-Gluc) convert HMR 1826 (vide supra) to Dox and that the generated drug produces a strong by-stander effect in cell culture. In vivo functionality of the s-β-Glu/HMR system was assessed by comparing the sensitivity of established con JEG-3 (non-expressing control) and JEG-3sG xenografts with HMR 1826 and Dox. Only tumors established with JEG-3sG immediately started to regress. Comparing Dox and HMR 1826, a dramatic effect was observed with the prodrug, already at 100 mg/kg confirming the superiority of the prodrug-converting system.

Four years later, transduction of tumor cells to secrete a targeted form of β-glucuronidase was also reported by de Graaf et al. [74] who noted a pronounced antitumor efficacy of DOX-GA3 after such adenoviral vector-mediated expression of human glucuronidase (GUS). The adenovirus vector, designated Ad/C28-GUSh, encoding GUS was fused to a human single-chain Fv against the epithelial cell adhesion molecule (EpCAM). Growth inhibition of well-established FMa human ovarian cancer xenografts was significantly delayed after i.v. injection of DOX-GA3 versus placebo. This effect was more pronounced after intratumoral administration. Virus alone has no effect.

Nitroreductase gene-directed enzyme prodrug therapy led Hay et al. to design and synthesize nitroaryl carbamate prodrugs of Dox [75]. Drawing a lesson from their own experiments in GDEPT with other cytotoxic agents and from the fact that a 4-nitrobenzylcarbamate linked to the 3'-amine of the sugar moiety was not released upon reduction [76], they prepared two types of prodrugs including or not a self-immolative spacer (*p*-aminobenzylcarbamate) and a nitroimidazole or a *p*-nitrophenyl ring as reductive units (Fig. 12).

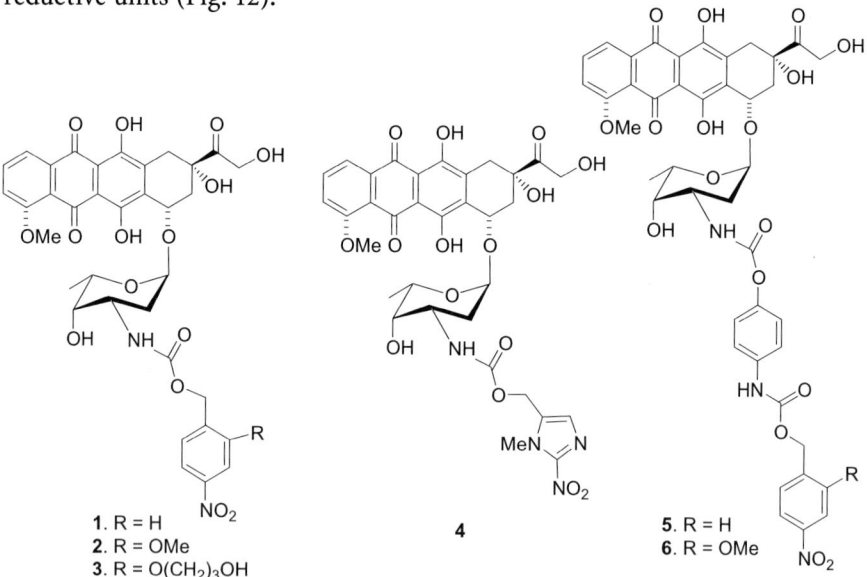

Fig. 12 Nitroreductase doxorubicin substrates

Prodrugs **1**, **2**, and **4** showed modest deactivation of Dox and relatively low selectivity for NTR. Whereas in the case of **4**, the addition of a spacer did not provide any significant increase in the deactivation of Dox or in selectivity for NTR+ve cells, in the nitrobenzyl analogues **1** and **2**, such addition affording **5** and **6**, did them. Prodrugs **5** and **6** were thus selected for in vivo evaluation. However, when evaluated against EMT6 tumors comprising c.a. 10% NTR+ve cells, no statically significant levels of killing, even for NTR+ve cells were observed. Moreover, the instability of **5** and **6** in culture medium along with the lack of in vivo activity represents a potential problem so authors concluded that optimization of pharmacokinetic and pharmacodynamic parameters were needed.

Self-immolative Dox and daunorubicin prodrugs for suicide gene therapy activation by carboxypeptidase G2 (CPG2) were prepared [77] by linking a glutamate residue to the 3'-amine of daunorubicin and doxorubicin. As a previous report [78] indicated that direct addition of L-glutamyl residue did not generate molecules that are substrates for CPG2, self-immolative spacers (i.e., p-amino or p-hydroxybenzyl carbamate) were inserted between these two units. The four conjugates **1–4** (Fig. 13) were stable and substrates for CPG2, **1** and **2** being more rapidly hydrolyzed than **3** and **4**. A significant decrease of cytotoxicity was displayed in the case of Dox prodrugs compared to that of daunorubicin prodrugs. Not only Dox-based prodrugs **2** and **4** were more deactivated (~20-fold versus ~7 fold) than the corresponding daunorubicin prodrugs **1** and **3**, but they yield higher cytotoxicity differential between control and cells expressing CPG2 intracellularly or extracellularly (1.4 to 23.3 versus 0.7 to 10.7 for **1** and **3**).

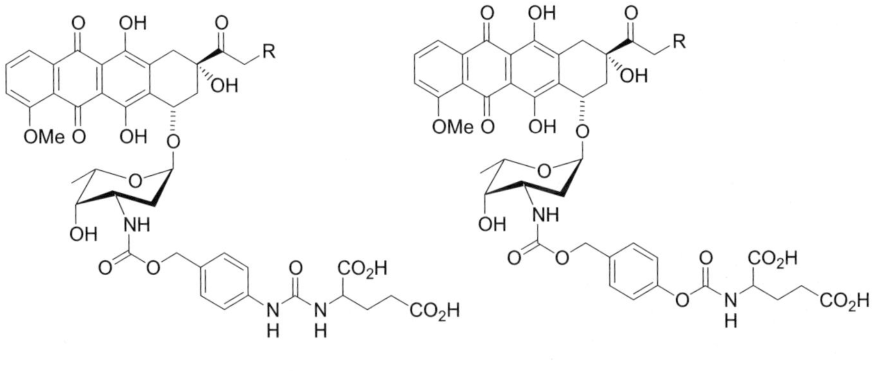

1. R = H
2. R = OH

3. R = H
4. R = OH

Fig. 13 Carboxypeptidase doxorubicin and daunorubicin substrates

2.3
Polymer-Directed Enzyme Prodrug Therapy (or PDEPT)

Most solid tumors possess pathophysiological characteristics such as extensive angiogenesis and hence hypervasculature, defective vascular architecture, and production of a number of permeability mediators. From impaired reticuloendothelial/lymphatic clearance results intratumor retention of macromolecular drugs thus delivered. All these characteristics, known as the enhanced permeability and retention (EPR effect), in macromolecular therapeutics have been reviewed by Maeda et al. [79]. On the other hand, as polymer applications for targeting of Dox are reviewed elsewhere by Kratz et al., only PDEPT will be briefly treated in this chapter.

Closely related to ADEPT, PDEPT, or polymer-directed enzyme prodrug therapy is a two-step antitumor approach in which the enzyme is linked to a polymer instead of a MAb, this polymer-enzyme conjugate generating cytotoxic drug outside the tumor cells, within the tumor interstitium through EPR effect.

The first PDEPT synthetic application in the field of doxorubicin was the report on the N-(2-hydroxypropyl)-methacrylamide copolymer doxorubicin or PK1, FCE28068, a conjugate (Fig. 14), which reached phase II clinical development [80, 81]. Following cellular uptake via pinocytosis, the linker of PK1 is cleaved by lysosomal enzymes, mainly cathepsin B.

From a practical point of view, in a first step, there is a systemic administration of a polymeric prodrug (MW ~30 000 g/mol) containing a linker designed for cleavage by the activating enzyme. Once the circulating polymer-drug conjugate has cleared (1–2 h), a polymer-enzyme can be administered as a second step. Having a higher molecular weight (~50 000–100 000 g/mol), this last conjugate circulated longer.

During a phase I study [82], PK1 demonstrated antitumor activity in refractory cancers, and no polymer-related toxicity. PK1 was about five-fold less toxic than Dox. This polymer conjugation, as expected, decreases doxorubicin dose-limiting toxicities. Next, using the model combination of PK1 and HMPA copolymer-cathepsin B, the feasibility of different aspects of the PDEPT concept was confirmed [83]. Thus these conjugates retain enzyme activity in vitro against both the low-molecular-weight and macromolecular substrates, increased circulation time in the blood until the passive targeting by the EPR effect was ascertained by using ^{125}I-Labelled cathepsin. In the B16F10 model it was showed that i.v. administration of PK1 produced a total doxorubicin AUC that was 17-fold higher than that seen for free Dox at equidose. Activity of the PDEPT combination was also controlled in COR-L23 tumor models.

Despite a large number of research studies, PK2 (FCE28069) is still the only targeted conjugate to be tested clinically [84]. Designed to recognize the hepatocyte asialoglycoprotein receptor, it has been explored for treating hepato-

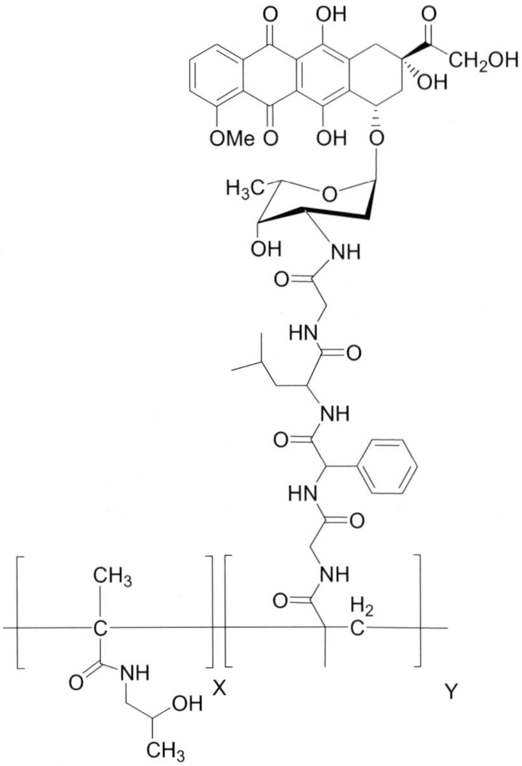

Fig. 14 PK 1 structure

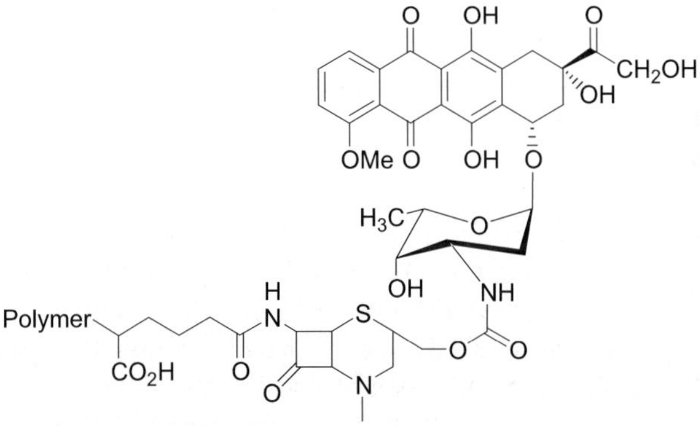

Fig. 15 Partial structure of HMPA-copolymer-β-lactame-Dox

cellular carcinoma. In phase I/II, the MTD was 160 mg/m^2 in Dox equivalent. The majority of the conjugate was present (gamma camera imaging) in normal liver with lower accumulations within hepatic tumor. Nevertheless, it was estimated that the concentration of this drug was still 12–50 fold higher than could be achieved with administration of free Dox [85].

More recently [86] a new model combination was investigated that included HMPA Copolymer-β-lactamase and HMPA Copolymer-C-Dox. Cephalosporine derivatives of doxorubicin have already been described as substrates of β-lactamases for the antibody-enzyme-prodrug-therapy [87]. Here, the potential disadvantage of β-lactamases that are not from mammalian origin and thus induce immunogenicity, should be counterbalanced by HMPA copolymer conjugate as conjugation of PEG and HMPA copolymers is well known to reduce immunogenicity of bound proteins.

2.4
Bacterial-Directed Enzyme Prodrug Therapy or BDEPT

This represents a new approach to use bacteria as vectors to deliver prodrug-releasing enzymes in tumor cells. In preliminary studies, Minton et al. [88] described the cloning of the gene-encoding nitroreductase enzyme into *Clostridium beijerinicki* NCIMB 8052. Following intratumoral injection into mice, localization of enzymatic activity to the tumor was observed. For their part, Vogelstein et al. created a strain of *Clostridium novyi* devoid of its lethal toxin called *Clostridium novyi-NT* [89]. When *C-novyi-NT* spores were administered together with cytotoxic drugs necrosis of tumors developed within 24 h, resulting in significant and prolonged antitumor effects [90]. They subsequently showed that such bacterial therapy can generate a potent immune response against experimental tumors [91], and then postulated that the bacterium's hemolytic properties could be exploited to enhance the release of liposome-encapsulated drugs within these tumors. To test this hypothesis with doxorubicin, they [92] treated syngeneic CT26 colorectal tumors in BALB/c mice by i.v. administration of *C. novyi-NT* spores and once germination has begun in the tumors, administered a single i.v. dose of liposomal Dox (Doxil). This combination resulted in complete regression of tumors in 100% of mice, 65% of the mice being still alive at 90 days. Identification of the liposome-disrupting factor led to a polypeptide, a neutral lipase encoded by the NT01CX2047 gene, originally called liposomase. As none of the commercially available lipase had significant liposome-disrupting activity, this means a specific activity of the NT01CX2047 lipase. Administration of *C. novyi-NT* spores plus Doxil resulted in increased drug concentration in the tumor without increasing in normal tissues. This effect was specific to *C. novyi-NT* and not the result of inflammation per se. This preclinical efficacy experiment, which was also similar with liposomes carrying CPT-11 (irinotecan) open the way to the possibility for liposomase to be attached to antibodies or to be encoded within vectors for gene-directed prodrug therapy.

2.5
Lectin-Directed Enzyme-Activated Prodrug Therapy or LEAPT

In this approach, the enzyme is targeted to the tumor cells by exploiting endogenous carbohydrate-lectin binding, followed by administration of a prodrug specifically activated by the targeted enzyme.

The asialoglycoprotein receptor or AGSPR is an endocytic lectin found in abundance on the surface of hepatocytes in the liver. In their synthetic application to deliver doxorubicin, Robinson et al. [93] used the cell-specific delivery of a synthetically glycosylated enzyme α-L-rhamnosidase (naringinase) and then a rhamnoside-capped prodrug. For this lectin-targeted enzyme, galactose was selected as the cell-specific ligand as well as the multivalent effect (10–15 carbohydrates per enzyme) exploited while combining enzymatic and chemical glycosylation. To establish the therapeutic usefulness of this strategy, intrasplenic injection of the human hepatocellular carcinoma HepG2 in nude mice was realized, followed after 20 min, by injection of prodrug Dox-Rha (10 mg/kg). Efficacy was demonstrated after 42 days of three-times-weekly dosing by quantification of the liver tumor burden. Total tumor burden and foci were significantly reduced compared with the control groups, galactosylated enzyme alone or Rha-Dox alone.

3
Tumor-Associated Enzymes Activation of Doxorubicin-Conjugates

Recently there has been increasing interest and achievement in the development of tumor-activated prodrugs which are based on the concept of utilizing endogenous enzymes that are consistently overexpressed in tumors to activate conjugates [94] similar to what has been done with exogenous enzymes.

3.1
Prostate-Specific Antigen or PSA

PSA is a serine protease with chymotrypsin-like activity that is a member of the kallikrein gene family [95]. PSA has been immunodetected in numerous human normal and tumor tissues [96] with breast and prostate tissues expressing the highest levels of PSA. However, PSA is relatively specific to prostate tissue, which synthesizes and secretes at least 100 times more PSA than breast tissue. PSA is used as a serological marker for the presence of prostate cancer. Serum PSA levels correlate not only with the existence of but also with the extent of prostate cancer; higher levels indicate a larger tumor burden, including metastatic disease. In the prostate gland, the mature form of PSA is enzymatically inactive because of the high concentration of zinc ion [97]. In addition, the proteolytic activity of secreted PSA is substantially reduced in the systemic circulation because of the formation of a covalent complex between PSA and the plasma protease inhibitors, 1-antichymotrypsin and 2-macroglobulin [98, 99]. Thus, secreted PSA is only enzymatically active in the microenvironment that surrounds prostate cells. The use of a prodrug being activated by PSA should therefore preferentially target PSA-secreting cells.

Researchers at Merck Research Laboratories [100] established a PSA cleavage map for human semenogelin created by digestion of semenogelin with human PSA and isolation of the digestion fragments. The peptide bond between Gln and Ser at positions 349 and 350 in semenogelin was the most readily cleaved peptide bond in this substrate. Then a systematic modification of the amino-acid residues surrounding this site led to the synthesis of a peptide of seven amino acid residues that was rapidly hydrolyzed by PSA. This peptide, N-glutaryl-(4-hydroxyprolyl)AlaSer-cyclohexaglycyl-GlnSerLeu-CO$_2$H was covalently linked to the aminoglycoside of doxorubicin giving the L-377,202 conjugate.

In vitro, L-377,202 was evaluated in cell culture and animal models of human prostate cancer cell growth and tumorigenesis. This compound has a greater than 20-fold selectivity against human prostate PSA-secreting LNCaP cells relative to the non-PSA-secreting DuPRO cell line.

In vivo, nude mouse xenograft studies showed reduced PSA levels by 95% and tumor weight by 87% at a dose below its MTD. Both free Dox and Leu-Dox were ineffective in reducing circulating PSA and tumor burden at their maximum tolerated doses. In contrast to mice with conventional treatment with doxorubicin, mice given L-377,202 showed a sharp reduction in average

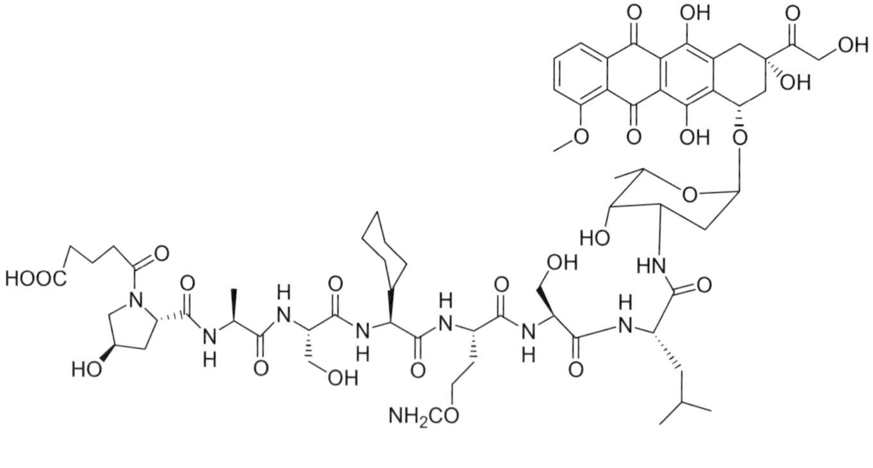

Glutaryl-Hyp-Ala-Ser-Chg-Gln-Ser-Leu-Dox

Fig. 16 Dox conjugate L-377,202 as substrate for PSA activation

tumor size. Overall, L-377,202 was about 15 times better than doxorubicin at inhibiting tumor growth in the mice. None of the mice treated with L-377,202 died.

Similar results were obtained using a second transplantable human prostate cancer tumor that secrets PSA, CWR22. In control studies, treatment of a non-PSA-producing cancer tumor xenograft (DuPro-1) with L-377,202 showed only a small amount of anti-tumor activity. As expected, a peptide-doxorubicin conjugate related to L-377,202, L-374,948 that was not cleaved by PSA was no more effective than conventional doxorubicin against PSA-secreting CWR22 tumors. L-377,202 was approximately 15 times more effective against LNCaP tumors than was doxorubicin. Moreover, L-377,202 did not show signs of gross toxicity at its maximally effective treatment doses.

In human [90], to evaluate safety and pharmacokinetics (PK), and determine the recommended dose for efficacy studies, 19 patients with advanced hormone-refractory prostate cancer were treated intravenously with 71 cycles of L-377,202 at escalating dose levels, once every 3 weeks. Toxicity, response and PK of L-377202 were assessed. It appeared that L-377202 was well tolerated. Dose-limiting grade 4 neutropenia was noted in two of two patients administered 315 mg/m^2 (both patients were able to resume therapy at 225 mg/m^2). The recommended dose for efficacy studies was 225 mg/m^2, which induced grade 4 neutropenia in one of six patients. Pharmacokinetics studies demonstrated that L-377202 was metabolized to Leu-Dox and Dox. PK were linear after administration of single doses of 225 mg/m^2, the mean area under the concentration-time profiles of L-377202, Leu-Dox, and Dox were 6 μmol·L/h, 4 μmol·L/h, and 1 μmol·L/h, and peak concentrations were 14 μmol/L, 5 μmol/L, and 120 nmol/L, respectively. At 225 and 315 mg/m^2, five patients completed at least three cycles of therapy; two patients had a greater than 75% decrease in PSA, and one patient had a stabilized PSA. No response was noted at dose levels less than 225 mg/m^2.

3.2
Plasmin

The plasmin system plays a key role in tumor invasion and metastasis by its matrix degrading activity and its involvement in tumor growth, in particular by its participation in growth factor activation and angiogenesis [102]. Active plasmin catalyzes the breakdown of extracellular matrix proteins and, thus, contributes to migration, invasion, and metastasis of tumor cells [103]. In the body, plasmin is predominantly present in its inactive pro-enzyme form plasminogen. Active plasmin is formed locally at or near the surface of tumor cells by urokinase-type plasminogen activator (u-PA) produced by the cancer and/or stroma cells. Plasmin activity is kept localized because cell-bound urokinase can activate cell-bound plasminogen into active plasmin, which stays cell-bound. Active urokinase and active plasmin do not occur in the blood circulation since they are very rapidly inhibited by inhibitors such as PAI-1 and α_2-antiplasmin, respectively. Many tumor cell lines and tumors have a significantly higher u-PA level than that of their normal counterparts, and u-PA was shown to be correlated with invasive behavior and to be a strong prognostic factor for reduced survival and increased relapse in many types of tumors [104–106]. Plasmin is a very promising enzyme for exploitation in a tumor-specific prodrug approach because the proteolytically active form is localized at tumor level.

The first anthracycline prodrugs that were designed for specific activation by plasmin were reported by Katzenellenbogen et al. [107]. However, the bipartate anthracycline prodrugs consisting of a tripeptide coupled to the 3'-amino sugar moiety appeared to be poor plasmin substrate because of steric factor as hypothesized [108, 109]. In a recent report [110], plasmin served as the target enzyme for tripartate prodrugs consisting of a tripeptide specifier (Fig. 17) coupled to the drug via self-immolative 1,6-elimination cyclization spacers.

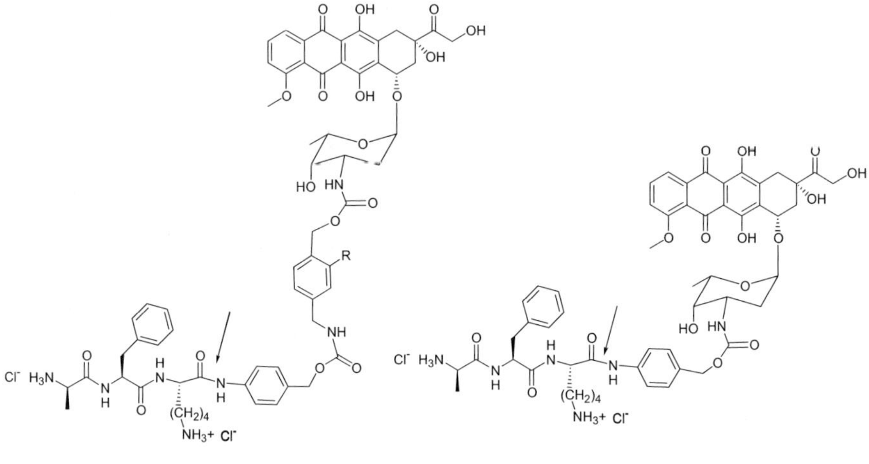

Fig. 17 Dox conjugates as plasmin substrates

Two compounds have been studied including a short spacer, ST-9802, and a longer one, ST-9905. Both prodrugs demonstrated selective cytotoxicity against plasminogen-activating cells in culture. A much higher cytotoxic potency was observed against murine EF43.fgf-4 cells, which produce high levels of uPA, versus MCF7 cells, which produce low levels of uPA. Confirming the implication of plasmin, an in vitro selective cytotoxicity of ST-9802 against uPA-transfected MCF-7 cells and not against parental cells was also observed, whereas prodrug toxicity against uPA-transfected cells markedly decreased in the presence of the plasmin inhibitor, aprotinin. The plasmin-mediated prodrug activation was confirmed by a markedly decreased cytotoxicity when aprotinin was added to EF43.fgf-4 cell cultures.

In vivo, no systemic toxic effects of ST-9802 and ST-9905 were displayed, in contrast to Dox. This was consistent with hydrolysis of prodrugs by plasmin mainly locally generated in the vicinity of tumor cells. The presence of physiological plasmin inhibitors in the systemic circulation would likely restrict exposure of the heart to active Dox, avoiding cardiotoxicity. The MTD of the ST-9905 derivative was more effective than the MTD of ST-9802 at equimolar concentration. Comparison of equimolar concentrations of ST-9905, ST-9802 and Dox in mice bearing EF43.fgf-4 or MCF7 tumors revealed that both prodrugs, in sharp contrast to Dox, displayed antiproliferative without discernible toxicity. Compound ST-9905 was cleaved more rapidly by plasmin than ST-9802, displaying a significantly greater antitumor efficacy than ST-9802 in EF43.fgf-4 tumors (68% of tumor inhibition with ST-9905 vs. 43% with ST-9802).

This marked in vivo antitumor efficacy of prodrugs was associated with strong inhibition of angiogenesis, as determined by vessel density assays and immunostaining with anti-vWF, anti-PECAM, or anti-type IV collagen antibodies [111].

3.3
Matrix Metalloproteinases

Matrix metalloproteinases (MMPs) are a family of Zn-dependent enzymes (> 20) that are synthesized as inactive proenzymes to become activated by proteolysis. MMPs have functions essentials for tumor formation, invasion and progression to the metastatic phenotype [112] For instance, the secreted enzyme pro-MMP-2 is proteolyzed and activated by the membrane-anchored MMP-14 [113]. Once activated, MMPs can cleave a variety of extracellular matrix proteins, such as collagen, laminin, fibronectin, and elastin, which are potentially important for tumor angiogenesis, invasion, and metastasis. Additionally, MMPs cleave a variety of other proteins, such as growth factor receptors and cell-adhesion molecules, which may be important for tumor growth and survival. The increased MMP activity in tumorogenic as compared to normal processes has made theses proteases attractive targets for inhibitors. MMP-2 (gelatinase A), and MMP-9 (gelatinase B) share many characteristics in their activation and regulation pathway. They must be localized on specific cell-surface receptors to be activated. They are associated with both the early and later stage in tumor progression and are particularly recognized for their roles in invasiveness in a variety of solid tumors and thus represented interesting targets for fighting cancer.

P. Senter et al. from Seattle Genetics [114] reported on the capacities of these proteases to be potentially good candidates for enzyme prodrug monotherapy. The design Dox prodrug was obtained by N-acylation of the amino-sugar moiety with the MMP substrate sequence, acetyl-L-prolyl-L-Leucyl-glycyl-L-Leucine, which by occupation of the P3–P1' substrate sites should be cleavable at the scissile Gly-Leu bond by the metalloproteinase MMP-2 and MMP-9.

The prodrug (Fig. 18) was significantly deactivated (30 times less toxic than Leu-Dox) and hydrolyzed at the predicted Gly-Leu bond to restore the fully active Leucyl Dox with a complete proteolysis within 4–16 h with MMP-9 protease but with a longer time with MMP-2. However these encouraging results were obtained with enzyme concentration and exposure time that are likely not to be physiologically relevant. Unfortunately, in a preliminary in vitro cytotoxicity experiment, authors were unable to detect any differences in the sensitivities of HT-1080 human fibrosarcoma cell line (expressing MMP-2 and MMP-9) and MB-MDA-453 (not expressing). Therefore, prodrugs need to be structurally optimized.

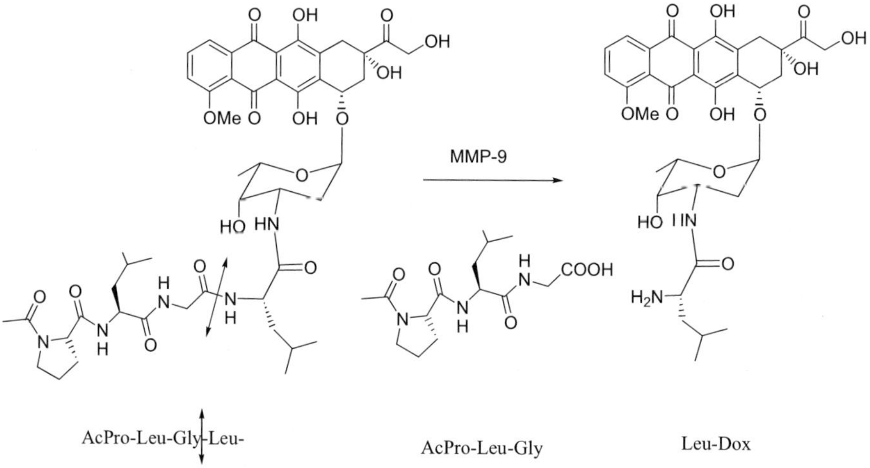

Fig. 18 Dox conjugate as substrate for MMP9 from [114]

A researcher from Bristol-Myers Squibb also recently reported (MMP)-activated prodrugs formed by coupling MMP-cleavable peptides to doxorubicin [115]. The resulting conjugates, such as that represented in Fig. 19, were excellent in vitro substrates for MMP-2,-9, and -14. The same fibrosarcoma cell line HT1080 as above, was used as a model system to test these prodrugs as these cells. In cultured HT1080 cells, simple MMP-cleavable peptides are known to be primarily metabolized by neprilysin, a membrane-bound metalloproteinase. Finally, MMP-selective metabolism in cultured HT1080 cells was obtained by designing conjugates that were good MMP substrates but poor neprilysin substrates.

Metabolization studies showed that MMP-selective conjugates were preferentially metabolized in HT1080 xenografts, relative to heart and plasma,

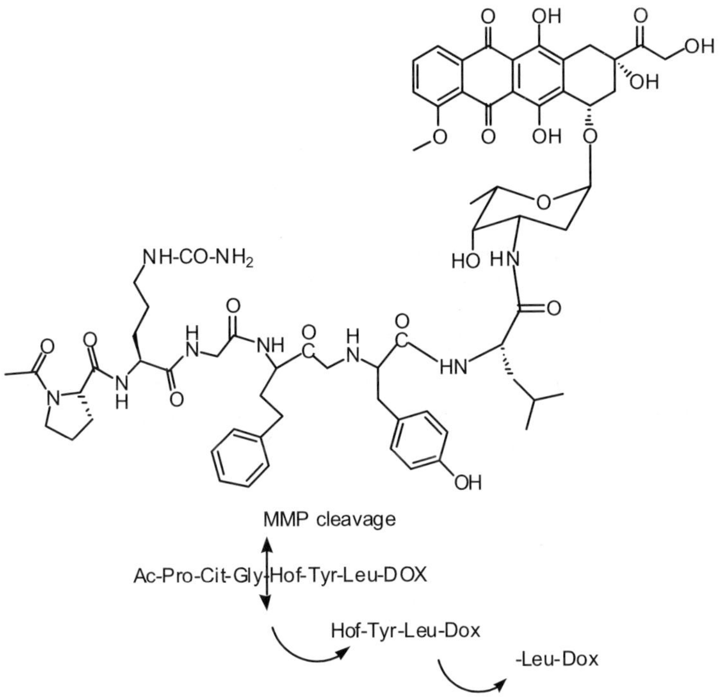

Fig. 19 Dox conjugate as substrate for MMP9 from [115]

leading to 10-fold increases in the tumor/heart ratio of doxorubicin. The doxorubicin deposited by a MMP-selective prodrug was more effective than doxorubicin at reducing HT1080 xenograft growth. In particular, this compound cured 8 of 10 mice with HT1080 xenografts at doses below the maximum tolerated dose, whereas doxorubicin cured 2 of 20 mice at its maximum tolerated dose. It appeared less toxic than doxorubicin at this efficacious dose because mice treated with compound do not show detectable changes in body weight or reticulocytes, a marker for marrow toxicity. Hence, MMP-activated doxorubicin prodrugs possess a much higher therapeutic index than doxorubicin using HT1080 xenografts as a preclinical model. Mouse tumors are poor models for human tumors with respect to MMP expression and activity. There is a lack of correlation between the in vitro enzymatic efficiencies for conjugate cleavage and the metabolism rates in cultured HT1080 cells, which is unclear. The MMP-activated prodrugs do not possess the optimal property of slowest rate of initial activation step and subsequent rapid steps compared with the initial step, for optimal prodrug performance. Optimization of the final release mechanism has to be done for further application in humans.

3.4
Endopeptidase Legumain

Legumain is a novel evolutionary offshoot of the C13 family of cysteine proteases. It is well conserved in plants and mammals, including humans. It is a robust acidic cysteine endopeptidase with remarkably restricted specificity, absolutely requiring an asparagine at the P1 site of its substrate sequence [116]. The selection of legumain as a target for tumor therapy was based on the fact that the gene encoding of this asparaginyl endopeptidase was found to be highly upregulated in many murine and human tumor tissues [117] but absent or present only at very low levels in all normal tissues from which such tumors arise. Importantly, overexpression of legumain occurs under such stress conditions as tumor hypoxia, which leads to increased tumor progression, angiogenesis, and metastasis.

The function of proteinases is to recognize a specific sequence (a stretch of peptide) in proteins and cleave to the proteins to either activate or destroy them. Tumors produce legumain to activate other proteinases, and together, these proteinases will digest the proteins that make up the surrounding tissues, and therefore make way for tumor cells and facilitate the spread of the tumor to nearby areas.

To take advantage of this activity of legumain, Liu et al. [118] designed a tumor activated prodrug of Dox for potential breast cancer adjuvant chemotherapy, called legubicin.

It has been shown that doxorubicin tolerates the addition of a leucine residue at this site. However, incorporation of additional amino acids abolishes cytotoxic activity [119]. Thus a prodrug analog such as N-(-t-Butoxycarbonyl-L-alanyl-L-alanyl-L-asparaginyl-L-leucyl) doxorubicin, was synthesized by the addition of an asparaginyl endopeptidase substrate peptide Boc-Ala-Ala-Asn-Leu to the amino group of doxorubicin through a peptide bond at COOH terminus of leucine. Upon cleavage by legumain, the prodrug is converted to a leucine-Dox derivative, thereby regaining cytotoxic function. The protective Boc group at the NH_2 terminus prevents aminopeptidase hydrolysis of the peptidyl component.

Legubicin led to tumor eradication in human breast cancer models with no toxicity compared to its parent drug, indicating that this prodrug has a good potential for developing anticancer drug.

First, the cytotoxic activity of legubicin [118] upon activation by legumain was analyzed in vitro using legumain expressing cell (+ 293 cell) and negative as control (– 293 cells). The effect of doxorubicin on both 293 cell types was similar, with legumain + cells only slightly more resistant to doxorubicin. In contrast, the cytotoxic effect of legubicin on control 293 cells was < 1% of that of doxorubicin, indicating peptide conjugation had abolished the cytotoxic effect of the doxorubicin. In contrast, a profound cytotoxic effect of legubicin was observed for legumain + 293 cells.

The in vivo effects of legubicin on normal and tumor-bearing hosts, and efficacy in tumor eradication were investigated using the CT26 murine syngeneic colon carcinoma model. Legubicin was very well tolerated in mice with much reduced toxicity compared with doxorubicin. I.p. injection of legubicin at 5 mg/kg three times at 2-day intervals induced complete growth arrest of the tumors with little evidence of toxicity, as most readily evidenced by the absence of weight loss. In contrast, doxorubicin failed to produce similar antitumor efficacy at doses approaching its maximum-tolerable dose. When doxorubicin was administered by the same protocol and dosage as for legubicin, toxicity was fatal.

A single injection of 5 mg/kg legubicin histologically induced more profound tumoricidal effects than animals given a comparable dose of doxorubicin. TUNEL assay analysis of tumor tissues revealed a higher apoptotic index for legubicin than for doxorubicin treatment. Surprisingly, in organs that do express legumain, such as kidney and liver, no injury was evident. These observations indicate that legubicin has significantly improved safety and therapeutic indices compared with doxorubicin.

3.4.1
Extracellular THIMET Oligopeptidase (TOP)

Trouet et al., rather than to target known peptidase specific of a given tumor, preferred to use a more empirical approach, the concept of extracellularly tumor activated prodrug (ETAP). On the basis of previous experience with L-Dox [120], they have developed a new compound CPI-004Na [121–123], unable to enter cell, stable in plasma, but which could be activated by enzymes specifically released from solid tumor cells. From the screen of a library of small peptides, a tetrapeptide N-β-alanyl-L-leucyl-L-alanyl-L-leucyl-Dox as peptide moiety was identified (Fig. 20).

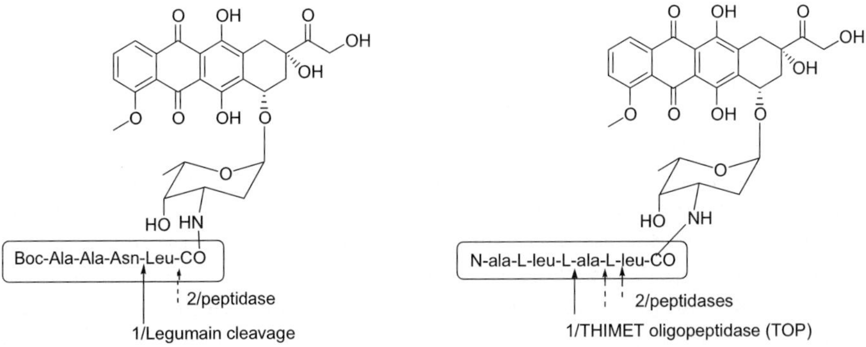

Fig. 20 Dox conjugates activation with endopeptidase legumain and THIMET oligopeptidase

The β-alanyl residue in the first position was chosen to provide blood stability. Hence, after 1 h, when 99% of N-alanyl-L-leucyl-L-alanyl-L-leucyl-Dox was degraded, only 10% of degradation was observed with N-β-alanyl-L-leucyl-L-alanyl-L-leucyl-Dox. The later is only 25% degraded after 7 h.

Upon incubation with MCF-7/6 cells, the conjugate is rapidly cleaved to yield dipeptidyl derivative Ala-Leu-Dox that gives, later on, Leu-Dox, known to be cleaved intracellularly by peptidase. Thus, a two-step extracellular activation is processing.

In vivo, two breast tumors MCF-7/6 and MAXF-1162 were studied. A reduced in vivo toxicity was observed of CPI-004Na as compared with Dox. The LD50 was 100 μM/kg and 155 μM/kg by i.p. route. The range of toxicity between CPI and Dox is 3.3 to 6.9 fold by i.v and i.p route, respectively. CPI-004Na is more active than Dox in human tumor Xenograft models. In mice bearing established tumor xenograft, an inhibition of tumor growth by 63% on day 63 at 34.5 μmole/kg was observed. From tissue distribution and pharmacokinetics studies the AUCs obtained for all tissues are much lower than after Dox.HCL treatment and importantly decreased, ranging from 80% for lung to 93% in the heart. In particular the heart exposure is reduced > 10-fold. The AUC in mice tumors treated with CPI-004Na is almost doubled (192%) vs. Dox.HCl. A relative high dose concentration was observed in kidney, but it is known that nephrotoxicity is higher in rodent than in human. In 2006, the identity of the enzyme responsible for the first rate-limiting step of CPI-004Na activation was initially attributed to a 70 KDa acidic (pI = 5.2) thiol dependant metallopeptidase further characterized as Thimet oligopeptidase (TOP). Interestingly, the activity of the enzyme is inhibited in oxygenated media such as blood and enhanced in mildly reducing and anoxic environments that are often characteristic of solid tumor.

4
Receptor-Mediated Targeted Delivery

4.1
Folate Receptor

Many types of cancer cells have a great affinity for folate—a form of water-soluble B vitamin—because they need the nutrient in order to grow and divide. The vitamin folic acid is a ligand capable of targeting covalently attached bioactive agents, quite specifically, to folate-receptor (FR) positive tumor cells. Thus, folate-targeted drug delivery has emerged as an alternative therapy for the treatment and imaging of many cancers and inflammatory diseases. Due to its small molecular size and high and specific binding affinity for cell surface folate receptors (FR), folate conjugate have the ability to deliver a variety of molecular complexes to pathologic cells without causing harm to normal tissues, and the number of molecules internalized can be very large ($> 10^6$ per h). Moreover, the FR, a tumor-associated protein, has been detected at high concentration in more than 90% of ovarian and other gynecological cancers. FR can actively internalize bound-folates via endocytosis and is expressed at different levels in other cancers such as in kidney, brain, lung, and breast carcinomas [124, 125]. A review summarize the applications of folic acid as a targeting ligand and highlight the various methods being developed for delivery of therapeutic and imaging agents to FR-expressing cells [126].

A folate-targeted biodegradable polymeric micellar system was developed [127] for doxorubicin. Thus, Dox and folic acid (FOL) were separately conjugated to a di-block copolymer of poly(D,L-lactic-coglycolic acid)-polyethylene glycol) (PLGA-PEG). Next, Dox-PLGA-PEG and PLGA-PEG-FOL were mixed with deprotonated Dox under a basic condition to produce mixed Dox micelles entrapping Dox aggregates within the core while exposing FOL ion the surface. These folate-targeted polymeric micelles exhibited enhanced and selective targeting against folate-positive cancer cells in vitro. In vivo, animal study also showed significant tumor-suppression effect for a human tumor xenograft nude mouse model.

In a following report [128], FOL and Dox were separately conjugated at α and ω terminal end of a PEG chain to produce FOL-PEG-Dox. The authors hypothesized that this linear and flexible FOL-PEG-Dox could sterically stabilize deprotonated and hydrophobic Dox nanoaggregates in an aqueous solution by anchoring the conjugated Dox moiety to Dox aggregates while exposing the more hydrophilic FOL moiety outside.

Folate-targeted of liposomal nanocarriers loaded with Dox have been reported several times. One pioneering work is that of Goren et al. [129], who attached folic acid to a certain number of PEG tether terminals of Doxil® through an amide bond. Rapid internalization of liposomes into FR-positive cells followed by drug release in the cytoplasmic compartment was observed by confocal fluorescence microscopy. The in vitro cytotoxicity of the liposome conjugated to folic acid was ten-fold stronger versus Doxil values and the same as free doxorubicin. FR-targeted liposomal doxorubicin was more efficient compared to doxorubicin in an in vivo assay against M109-HiFR tumor implanted in BALB/c mice. Later on [130], the same group investigated the uptake of folate-targeted liposomal carriers in the J6456 lymphoma tumor model up-regulated for the folate receptor and found that the drug levels in ascetic J6456-FR increased by 17-fold, whereas those in plasma decreased by 14-fold when compared with non-targeted experiments. They also discussed the pros and cons of the liposome platform using a liposome drug delivery system targeted to the folate receptor [131] as an example.

The potential of folate-receptor mediated liposomal delivery of liposomes loaded with doxorubicin was again illustrated by several recent reports [132–134].

An erythrocyte-based delivery system represents a natural, biocompatible, and non-immunogenic novel drug carrier that provides optimized blood concentration by protecting the drug from metabolism and by controlling the kinetics of drug release. Interestingly, these vesicles should also reduce heart toxicity. Thus, Mishra and Jain proposed [135] to load these vesicles with doxorubicin and to coat them with folic acid derivative. The stability and decrease cytotoxicity of these Dox-Fv associated with the increase in life of Balb/c mice bearing murine leukaemia L1210 and the lack of undesirable effects made them interesting.

Plant viruses that have genomic materials encapsidated within protein cages may provide viable delivery platforms for drugs, either by drug loading or by chemical conjugation. Using the Hibiscus chlorotic ringspot virus as a model plant virus, Ren et al. [136] prepared nanosized protein cages (30 nm) capable of encapsulating doxorubicin. Folic acid was conjugated onto the capsids and the resulting nanosized systems improved the uptake and cytotoxicity of doxorubicin in the ovarian cancer cells, OVCAR-3. However, further information on the immune response and in vivo efficacy remains to be cleared.

In the course of ADEPT research, N-phenylacetamido doxorubicin prodrugs, along with melphalan derivative, were proposed [137] to be selectively activated by monoclonal-antibody-penicillin-G amidase conjugates. As an alternative, instead of using monoclonal antibody, Zhang et al. synthesized [138] folate-conjugated penicillin G-amidase (folate-PGA), the folate : PGA molar ratio being approximately 4 : 1. Preliminary studies showed that more than 85% of specific activity of PGA remained and the activation rates for the conversion of prodrug to drug was the same. Other data such as internalization of the folate-PGA conjugates, complete release of Dox from its prodrug, in vitro antiproliferative effects, pharmacokinetics and biodistribution, were relevant and worthy of further investigation.

4.2
Somatostatin Receptor

Various primary human tumors as well as tumor cell lines were shown to possess somatostatin receptors (SSTRs). Experimental oncology studies demonstrated that somatostatin (SST) and its octapeptide analogues, such as octreotide or RC-160, can inhibit the growth of some malignancies. Most of the octapeptide SST analogues such as RC-160 and RC-121 bind selectively and with high affinity to SSTR2 and SSTR5 subtypes. In order to reduce the significant toxic effect of doxorubicin when used in the therapy of advanced androgen-independent prostate cancer, receptors for somatostatin (SST) have been used for targeting of doxorubicin on human prostate cancer specimen. Thus doxorubicin derivative, 2-pyrrolino-doxorubicin (AN-

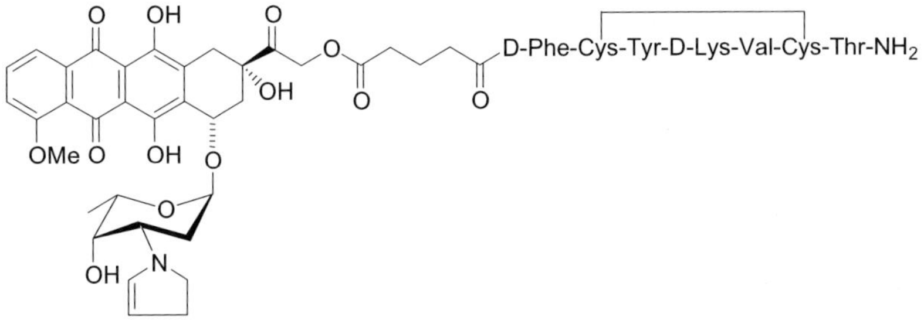

Fig. 21 Structure of AN-238, a pyrrolino-Dox

201) as the 14-O-hemiglutarate was linked to the amino terminus of RC-121 to give the conjugate, AN-238 [139]. First experiments showed that AN-238 inhibits growth of androgen-independent dunning R-3327-AT-1 prostate cancers in rats at non-toxic doses. Evidence was given that this conjugate binds strongly to SSTRs subtypes as RC-121 itself. More recent data [140] indicates that AN-238 inhibited powerfully the growth of endometrial carcinoma, which express SST receptors. Moreover, AN-238 caused a weaker induction of MDR-1 than the pyrrolino-doxorubicin in three cancer lines (HEC-1A, RL-95-2, and AN3CA) [141]. The same efficacy was found [142] in experimental non-Hodgkin's lymphomas, the human NHL cell lines RL and HT. SN-238 did not exhibit significant myelotoxicity, the most serious side-effect and the dose-limiting factor of cytotoxic drugs. Most of the side-effects, including loss of body weight or slight reduction of white blood cell count, were attributed to the high esterase activity in mouse serum, which can release the pyrrolidino-doxorubicin into the circulation. As this esterase activity is much lower in man, authors anticipated a lower toxicity of AN-238 in patients.

4.3
Bombesin Receptors

Bombesin-like peptides such as gastrin-releasing peptide (BN/GRP) have been identified as autocrine/paracrine growth factors in several human malignancies as well as binding sites for various bombesin receptor subtypes [143]. Consequently, the group of Schally developed a cytotoxic bombesin analogue, AN-215 [144]. As in the case of targeting somatostatin receptor, 2-pyrrolino-Dox or AN-201 was chosen as cytotoxic drug but was linked covalently, in the present case, to the bombesin analogue Gln-Trp-Ala-Val-Gly-His-Leu-Ψ-(CH$_2$-NH)-Leu-NH$_2$ (RC-3094). For studying efficacy and toxicity of AN-215, three in vivo renal cancer cell line models were used [145]. They showed that AN-215 significantly inhibited the growth of all these cell lines, whereas the pyrrolino doxorubicin alone had no marked effect. This efficacy was independent of the expression patterns of MDR-1 and MRP-1 in these RCC cell lines demonstrating that AN-215 can overcome chemoresistance in experimental cell carcinomas. The same observations were reported with experimental human breast cancers [146, 147] and human endometrial cancers [148]. Since human prostate cancers express a high level of receptors for bombesin/gastrin releasing peptide, AN-215 was also evaluated [149] in nude mice bearing subcutaneous xenografts of prostate cancers as DU-145, LuCaP-35, MDA-PCa-2b, or bearing intraosseous implants of C4-2 human prostate cancers.

4.4
Luteinizing Hormone Releasing Receptors (LHRH)

Based upon the fact that expression of LHRH receptors, a G-protein, coupled receptor has been demonstrated in various human cancers including breast, ovarian, endometrial, prostate and pancreatic, AN-152 was first synthesized as a conjugate of Dox and LHRH agonist, the [D-Lys^6]luteinizing hormone-releasing hormone (LHRH) [150]. The conjugate AN-207 was subsequently synthesized from 2-pyrrolino Dox derivative, AN-201. Another study was carried out in breast cancers [151] that express elevated levels of the HER-2 protein and belong to the ErbB/HER type I tyrosine kinase receptor family. These mammary cancers are resistant to chemothcrapy and have very poor prognosis. In this study, Schally's group investigated whether targeting Dox to LHRH-R, MX-1 estrogen-independent Dox-resistant breast carcinomas could improve the efficacy of treatment. Although AN-201, the 2-pyrrolino-Dox, which is noncross-resistant with Dox, showed a remarkable efficacy in MX-1 tumors, this efficacy was further improved with conjugate AN-152. The mechanism by which AN-152 can overcome the resistance of MX-1 tumors to Dox is not very clear, but among several hypothesis, it is more likely due to a higher concentration of the cytotoxic agent delivered to target tumor tissue.

The following three studies, which have been simultaneously reported by the same group, concern the targeting of human renal cell carcinomas, of non-Hodgkin's lymphomas, and of human malignant melanomas. In all these studies, AN-207 was involved. In the case of human renal cell carcinomas [152], positive staining for LHRH receptors was found in all of the 28 surgically removed specimens as well as in the three human RCC cell lines, A-498, ACHN, and 786-0. AN-207 significantly inhibited the growth of xenografts involving these cell lines, whereas 2-pyrrolino-Dox alone has poor effects. Blockade of LHRH receptors by an excess of antagonist Decapeptyl suppressed tumor inhibitory effects of AN-207.

Similar results were obtained in non-Hodgkin's lymphomas [153] and melanomas [154]. The targeting of cytotoxic luteinizing hormone-releasing hormone analogs to breast, ovarian, endometrial, and prostate cancers was recently reviewed [155].

4.5
Integrin Receptor

De Groot et al. [156] reported a relatively new approach for targeting drugs to receptors involved in tumor angiogenesis and metastasis. For potential enhanced tumor recognition potential, their approach consists of using a synergistic mode of action of two proteins to release active drug at the tumor site. They designed a doxorubicin prodrug that contains a dual tumor specific moiety incorporating a tumor-specific recognition site and a tumor selective enzymatic activation sequence. The first tumor-specific device was a bicyclic peptide CDCRGDCFC (RGD-4C) that selectively binds receptors $\alpha v \beta 3$ and $\alpha v \beta 5$ integrins. The later are known to be highly over-expressed on invading tumor endothelial cells (contact between cells and between endothelial and the extracellular matrix in metastasis) but also present in angiogenesis at the level of the endothelial cells in direct contact with bloodstream which control microcirculation within tumor mass. The second tumor-specific sequence was a D-ala-Phe-Lys tripeptide that is selectively recognized by the tumor-associated protease plasmin (component of the urokinase-type plasminogen activator system) which is involved in tumor invasion and metastasis.

The tumor-homing bis-disulfide containing RGD-4C peptide (CDCRGD-CFC) allow the selective binding to avb3 receptor on HUVECs by the prodrug represented in Fig. 22, with IC_{50} for integrin ligand between 25 nM (radio-active binding assay) and 150 nM (endothelial cell binding assay). It possessed plasmin substrate properties as investigated by in vitro incubation with plasmin. However, incomplete activation of the prodrug was observed. Additional proof of principle for plasmin cleavage was delivered in an in vitro cytotoxicity experiment. In the presence of the plasmin, the prodrug

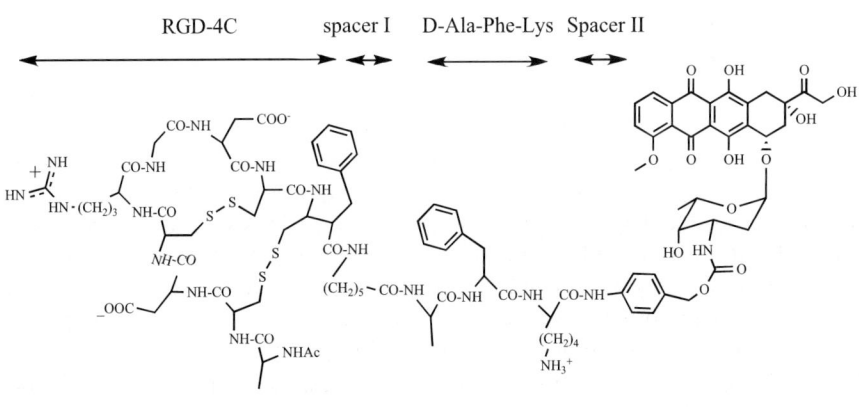

CDCRGDCFC = RGD-4C

Fig. 22 RGD integrin targeting associated with plasmin drug release

displayed a cytotoxic effect on fibrosarcoma HT1080 and HUVECs cells, approaching that of free Dox. Thus, if the proof of principle has been demonstrated, improvement was still needed, principally because of the low solubility of the conjugate.

Drug-delivery systems to achieve controlled release or enable drug targeting to specific tumor sites using sterically stabilized liposomes (SSL) were a major breakthrough in prolonging circulation time and achieving improved tumor targeting [157]. It has been demonstrated that SSL can accumulate in tumor tissue due to the effect of enhanced permeability and retention (EPR) [158–160]. The problem is that anticancer drugs accumulation in tumor tissue via SSL seems to be a prerequisite but far from sufficient to guarantee a therapeutic improvement. While introducing PEG enables liposomes to accumulate in tumor tissue, it creates a steric barrier that could cause a reduction in liposome interaction with the target cells and leads to low uptake of the entrapped drugs via cell endocytosis or membrane fusion [161]. This discrepancy of SSL accumulation in tumor tissue and normal tissue led to hypothesize that it is possible to enhance the intracellular delivery of the entrapped drugs accumulated in tumor tissue to obtain an improved therapeutic efficacy, in particular via passive targeting of angiogenesis using overexpressed receptors such as avβ3 integrin receptor. The RGD (arginine–glycine–aspartic acid) sequence is known to serve as a recognition motif in multiple ligands for several different integrins such as α_v-β_3 integrin and α_5-β_1 integrin [162].

The use of RGD peptides with affinity for this integrin coupled to the distal end of poly(ethylene glycol)-coated long-circulating liposomes (LCL) to obtain a stable long-circulating drug-delivery system functioning as a platform for multivalent interaction with α_v-β_3 integrins was reported by Storm et al. [163]. A charge of 300 RGD peptides has been determined on the surface of one liposome, based on the estimation that 80 000 phospholipid molecules form one liposome vesicle of 100 nm. Moreover, Dox content of liposomes was determined to be 80–150 mg Dox/mmol lipid.

The biological results show that cyclic RGD-peptide-modified LCL exhibited increased binding to endothelial cells in vitro. Moreover, intravital microscopy demonstrated a specific interaction of these liposomes with tumor vasculature, a characteristic not observed for LCL. RGD–LCL encapsulating doxorubicin inhibited tumor growth in a doxorubicin-insensitive murine C26 colon carcinoma model, whereas doxorubicin in LCL failed to decelerate tumor growth to angiogenic endothelial cells in vitro and in vivo.

It appeared that coupling of RGD to LCL targets redirected these liposomes to angiogenic endothelial cells in vitro and in vivo. Dox-containing RGD–LCL were able to decelerate tumor growth in a Dox-insensitive tumor model. In this model, LCL-encapsulated Dox failed to inhibit tumor growth. Likely, the superior therapeutic efficacy of RGD–LCL is the result of inhibition of tumor progression via inhibition of angiogenesis rather than via direct cytotoxic effects on tumor cells.

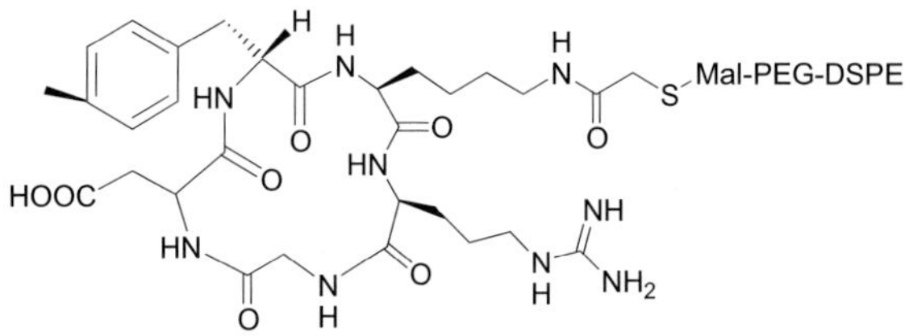

Fig. 23 Doxorubicin-loaded nanoparticles RGD-targeted

Recently, Zhang et al. [164] reported the use of RGD-mimetic-modified SSL (RGDm-SSL) aimed to achieve tumor accumulation as well as enhanced intracellular delivery loaded with doxorubicin. Flow cytometry and confocal microscopy reveal that RGDm-SSL facilitated the Dox uptake into the melanoma cells via integrin-mediated endocytosis. Dox-loaded RGDm-SSL (RGDm-SSL-Dox) displayed higher cytotoxicity on melanoma cells than Dox-loaded SSL (SSL-Dox). Tissue distribution and therapeutic experiments were examined in C57BL/6 mice carrying melanoma B16 tumors. RGDm-SSL-Dox displayed similar drug accumulation in tumor tissue to that of SSL-Dox but showed significantly lower Dox level in blood and remarkably higher Dox level in spleen than SSL-Dox. Administration of RGDm-SSL-Dox at a dose of 5 mg Dox/kg resulted in effective retardation of tumor growth and prolonged survival times compared with SSL-Dox. These results suggest that RGDm-modified SSL may be a promising intracellular targeting carrier for efficient delivery of chemotherapeutic agents into tumor cells.

Carbohydrate-based NucleoParticules (NPs) for the treatment of cancer have been reported by Bibbya et al. [165]. The NPs core was composed of cross-linked carbohydrate inulin multi-methacrylate (IMMA) with the cyclic RGD sequence cyclo(–Arg–Gly–Asp–d–Phe–Cys–), covalently attached to the NPs via PEG-400. 72% of the Dox were attached to the NP matrix via an amide bond, the remaining 28% being entrapped as unconjugated drug. Pharmacokinetics of these hydrophilic NPs allowed the determination of total, unconjugated and metabolized doxorubicin examined for 5 days following i.v. administration of the NP formulation (250 mg doxorubicin equiv.). Among several observations, unconjugated doxorubicin (and doxorubicin metabolites such as doxorubicinol) content was low in the liver, suggesting that upon distribution to this organ, the NPs are not readily metabolized. Exposure of doxorubicin to cardiac tissue was low.

5
Peptide Drug Delivery

Vectorization of Dox with peptide vectors could significantly reduce its accumulation in the heart [166]. On the other hand, to overcome the problem of transportation of Dox in the brain, the technology Pep:trans has been developed [167]. Such a strategy was based on short natural-derived peptides that are able to cross the blood–brain barrier without comprising its integrity. Thus, Dox was covalently linked to small peptide vectors, L-SynB1 of 18 amino acids or L-SynB3 of 10 amino acids and its enantiomer D-SynB3. Targeting of these conjugates was studied by in situ mouse brain perfusion method. This significantly increased the brain uptake of Dox (about 30-fold). The mechanism of this transport uses a saturable transport mechanism to cross the BBB, occurring via an adsorptive-mediated endocytosis. It must be noted that the effect of poly(L-lysine) and protamine, endocytosis inhibitors, reduced the brain uptake in a dose-dependent manner.

In another approach, the same group [168] tested the capability of two peptide vectors to deliver Dox in Pgp expressing cells, responsible for MDR. The Pegelin (such as SynB1) and Penetratin peptides as well as a 16-amino-acid-long peptide were linked to the 3′-amine group of Dox. The cytotoxic effect of these conjugates was measured in human erythroleukemic (K562/ADR) resistant cell line. Different experiments showed that the conjugates by-pass the Pgp. This was also true in an in situ brain perfusion method.

R = GKRKKKGKLGKKRPRSRC

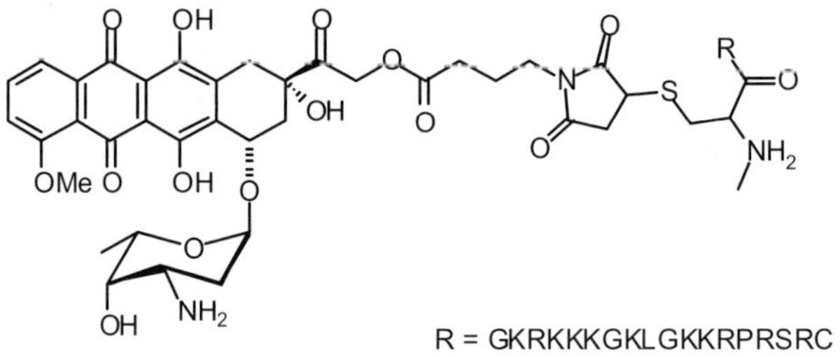

Fig. 24 Vectocell peptide conjugate

Conjugation of Dox to short (15–23 amino acids) peptide sequences called Vectocell peptides led to conjugates in which the in vivo therapeutic index of Dox is improved [169]. Three different types of Dox conjugates covalently attached to Vectocell peptides, polycationic cell-penetrating peptides, have been synthesized. Ester or thioether bonds have been used to attach the peptide at C-14 and an amide bond in the case of linking at the 3-amine group. Best results were obtained for both Dox-sensitive and -resistant cell models, when the specific Vectocell peptide (DPV 1047) was linked to the C-14 of Dox through an ester linker with $IC_{50} = 6\,\mu M$ for HCT116 and 133 μM in MCF7/Adr-resistant cell line versus > 1000 with Dox. Extensive evaluation of this conjugate was next undertaken and in vivo improved efficacy in MDA-MB 231 human breast adenocarcinoma and in colon model (HCT116) was found compared to Dox. This was also true in the HTC 15 human colorectal adenocarcinoma constitutively expressing P-glycoprotein.

References

1. Trail PA, King HD, Dubowchik GM (2003) Cancer Immunol Immunother 52:328
2. Wu AM, Senter PD (2005) Nat Biotechnol 23:1137
3. Hamann PR (2005) Expert Opin Ther Patents 15:1087
4. Trail PA, Willner D, Lasch JS, Henderson AJ, Hofstead SJ, Casazza AM, Firestone RA, Hellström I, Helström KE (1993) Science 261:212
5. Saleh MN, Sugarman S, Murray J, Ostroff JB, Healey D, Jones D, Daniel CR, LeBherz D, Brewer H, Onetto N, LoBuglio AF (2000) J Clin Oncol 18:2282
6. Tolcher AW, Sugarman S, Gelmon KA, Cohen R, Saleh H, Isaacs C, Young L, Healey D, Onetto N, Slichenmyer W (1999) J Clin Oncol 17:478
7. Trail PA, Willner D, Bianchi AB, Henderson AJ, Mark D, TrailSmith MD, Girit E, Lasch S, Hellström I, Hellström KE (1999) Clin Cancer Res 5:3632
8. Smith S (2001) Curr Opin Mol Ther 3:295
9. Wall AF, Donaldson KL, Mixan BJ, Trail PA, Siegall CB (2001) Int J Cancer 93:590
10. Ross HJ, Hart LL, Swanson PM, Rarick MU, Figlin RA, Jacobs AD, McCune DE, Rosenberg AH, Baron AD, Grove LE, Thorn MD, Miller DM, Drachman JG, Rudin CM (2006) Lung Cancer 54:69
11. Seattle Genetics Provides Update on SGN-15 Phase II. Clinical Program at the 11th World Conference on Lung Cancer, July 3–6, 2005, Barcelona, Spain
12. Muldoon LL, Neuwelt EA (2003) J Neurooncol 65:49
13. King HD, Staab AJ, Pham-Kaplita K, Yurgaitis D, Firestone RA, Lasch SJ, Trail PA (2003) Bioorg Med Chem Lett 13:2119
14. King HD, Yurgaitis D, Willner D, Firestone RA, Yang MB, Lasch SJ, Hellström KE, Trail PA (1999) Bioconjugate Chem 10:279
15. King HD, Dubowchik GM, Mastalerz H, Willner D, Hofstead SJ, Firestone RA, Lasch SJ, Trail PA (2002) J Med Chem 45:4336
16. Dubowchik GM, Radia S, Mastalerz H, Walker MA, Firestone RA, King HD, Hofstead SJ, Willner D, Lasch SJ, Trail PA (2002) Bioorg Med Chem Lett 12:1529
17. Acton EM, Tong GL, Mosher CW, Wolgemuth RL (1984) J Med Chem 27:638
18. Cherif A, Farquhar DJ (1992) J Med Chem 35:3208
19. Griffiths GL, Mattes MJ, Stein R, Govindan SV, Horak ID, Hansen HJ, Goldenberg DM (2003) Clin Cancer Res 9:6567
20. Sapra P, Stein R, Pickett J, Qu Z, Govindan SV, Cardillo TM, Hansen HJ, Horak ID, Griffiths GL, Goldenberg DM (2005) Clin Cancer Res 11:5257
21. Hebert C, Norris K, Sauk JJ (2003) J Drug Target 11:101
22. Inoh K, Muramatsu H, Torii S, Ikematsu S, Oda M, Kumai H, Sakuma S, Inui T, Kimura T, Muramatsu T (2006) Jpn J Clin Oncol 36:207
23. Vega J, Ke S, Fan Z, Wallace S, Charsangavej C, Li C (2003) Pharm Res 20:826
24. Shiah JG, Sun Y, Kopeckova P, Peterson CM, Straight RC, Kopecek J (2001) J Control Release 74:249
25. Bagshawe KD (1989) Brit J Cancer 60:275
26. Senter P, Springer C (2001) Adv Drug Deliv Rev 53:247
27. Bagshawe KD, Sharma SK, Begent RH (2004) Expert Opin Biol Ther 4:1777
28. Rooseboom M, Commandeur JNM, Vermeulen NPE (2004) Pharmacol Rev 56:53
29. Florent JC, Dong X, Gaudel G, Mitaku S, Monneret C, Gesson JP, Jacquesy JC, Mondon M, Renoux B, Andrianomenjanahary S, Michel S, Koch M, Tillequin F, Gerken M, Czech J, Straub R, Bosslet K (1998) J Med Chem 41:3572
30. Bosslet K, Czech J, Lorenz P, Sedlacek HH, Schuermann M, Seeman G (1992) Brit J Cancer 65:234

31. Gerken M, Krause M, Czech J, Bosslet K, Seemann G, Hoffmann D, Sedlacek HH (1991) Eur Patent 91101096.5
32. Bosslet K, Czech J, Hoffmann D (1994) Cancer Res 54:2151
33. Platel D, Bonoron-Adèle S, Dix RK, Robert J (1999) Brit J Cancer 81:24
34. Fishman WH, Anlyan MJ (1947) J Biol Chem 169:449
35. Connors TA, Whisson ME (1966) Nature 210:866
36. Double JA, Workman P (1977) Cancer Treat Rep 61:909
37. Bosslet K, Czech J, Hoffmann D (1996) Tumor Target 1:45
38. Bosslet K, Straub R, Blumrich M, Czech J, Gerken M, Sperker B, Kroemer HK, Gesson JP, Koch M, Monneret C (1998) Cancer Res 58:1195
39. Murdter TE, Sperker B, Kivistö KT, McClellan M, Fritz P, Friedel G, Linder A, Bosslet K, Toomes H, Dierkesmann R, Kroemer HK (1997) Cancer Res 57:2440
40. Murdter TE, Friedel G, Backman JT, McClellan M, Schick M, Gerken M, Bosslet K, Fritz P, Toomes H, Kroemer HK, Sperker B (2002) J Pharmacol Exp Ther 301:223
41. Woessner R, An Z, Li X, Hoffman RM, Dix R, Bitonti A (2000) Anticancer Res 20:2289
42. Sperker B, Werner U, Murdter TE, Tekkaya C, Fritz P, Rainer W, Adam U, Gerken M, Drewelow B, Kroemer HK (2000) Naunyn-Schmiedeberg's Arch Pharmacol 362:110
43. Sperker B, Werner U, Murdter TE, Tekkaya C, Fritz P, Rainer W, Adam U, Gerken M, Drewelow B, Kroemer HK (2000) Chem Abstr 133:271539v
44. Leenders RTGG, Gerritz KAA, Ruijtenbeek R, Scheeren HW, Haisma HJ, Boven E (1995) Tetrahedron Lett 26:1701
45. Leenders RTGG, Damen EWP, Bijsterweld EJA, Scheeren HW, Houba PHJ, van der Meulen-Muilcman IH, Boven E, Haisma HJ (1999) Bioorg Med Chem 7:1597
46. de Groot FMH, Damen EWP, Scheeren HW (2001) Curr Med Chem 8:1093
47. Houba PHJ, Boven E, Haisma HJ (1996) Bioconj Chem 7:606
48. Houba PHJ, Leenders RGG, Boven E, Scheeren JW, Pinedo HM, Haisma HJ (1996) Biochem Pharmacol 52:455
49. Houba PHJ, Boven E, Van Der Meulen-Muileman IH, Leenders RGG, Scheeren JW, Pinedo HM, Haisma HJ (2001) Int J Cancer 91:550
50. De Graaf M, Boven E, Oosterhoff D, van der Meulen-Muileman IH, Huls GA, Gerritsen WR, Haisma HJ, Pinedo HM (2002) Brit J Cancer 86:811
51. de Graaf M, Nevalainen TJ, Scheeren HW, Pinedo HM, Haisma HJ, Boven E (2004) Biochem Pharmacol 68:2273
52. Houba PHJ, Boven E, Erkelen CAM, Leenders RGG, Scheeren JW, Pinedo HM, Haisma HJ (1998) Brit J Cancer 78:1600
53. Vrudhula VM, Svensson HP, Senter PD (1995) J Med Chem 38:1380
54. Harding FA, Liu AD, Stickler M, Razo OJ, Chin R, Faravashi N, Viola W, Graycar T, Yeung VP, Aehle W, Meijer D, Wong S, Rashid MH, Valdes AM, Schellenberger V (2005) Mol Cancer Ther 4:1791
55. Alderson RF, Toki BE, Roberge M, Geng W, Basler J, Chin R, Liu A, Ueda R, Hodges D, Escandon E, Chen T, Kanavarioti T, Babé L, Senter PD, Fox JA, Schellenberger V (2006) Bioconjugate Chem 17:410
56. Andrianomenjanahary S, Dong X, Florent JC, Gaudel G, Gesson JP, Jacquesy JC, Koch M, Michel S, Mondon M, Monneret C, Petit P, Renoux B, Tillequin F (1992) Bioorg Med Chem Lett 2:1093
57. Gesson JP, Jacquesy JC, Mondon M, Petit P, Renoux B, Andrianomenjanahary S, Dufat-Trinh Van H, Koch M, Michel S, Tillequin F, Florent JC, Monneret C, Bosslet K, Czech J, Hoffmann D (1994) Anti-Cancer Drug Design 9:409

58. Gosh AK, Kan SR, Farquha RD (1999) Chem Commun 24:2527
59. Bakina E, Farquhar D (1999) Anti-Cancer Drug Design 14:505
60. Farqhar D, Cherif A, Bakina E, Nelson J (1998) J Med Chem 41:965
61. Torgov MY, Alley SC, Cerveny CG, Farquhar D, Senter PD (2005) Bioconjugate Chem 16:717
62. Heinis C, Alessi P, Neri D (2004) Biochem 43:6293
63. Gopin A, Pessah N, Shamis M, Rader C, Shabat D (2003) Angew Chem Int Ed 42:327
64. Amir RJ, Pessah N, Shamis M, Shabat D (2003) Angew Chem Int Ed 42:4494
65. Shabat D, Amir RJ, Gopin A, Pessah N, Shamis M (2004) Chem Eur 10:2626
66. Shamis M, Lode HN, Shabat D (2004) J Am Chem Soc 126:1726
67. Shabat D, Popkov M, Shamis M, Lerner RA, Barbas III CF, Shabat D (2005) Angew Chem Int Ed 44:718
68. de Groot FM, Albretch C, Koekkoek R, Beusker PH, Scheeren HW (2003) Angew Chem Int Ed 42:4490
69. Szalai ML, Kevwitch RM, McvGrath DV (2003) J Am Chem Soc 125:15688
70. de Groot FMH, Loos WJ, Koekkoek R, van Berkom LW, Busscher GF, Seelen AE, Albretch C, de Bruijn P, Scheeren HW (2001) J Org Chem 66:8815
71. Bridgewater G, Springer CJ, Knox R, Minton N, Michael P, Collins M (1995) Eur J Cancer 31A:2362
72. Niculescu-Duvaz D, Niculescu-Duvaz I, Friedlos F, Spooner R, Martin J, Marais R, Springer CJ (1999) J Med Chem 42:2485
73. Weyel D, Sedlacek HH, Müller R, Brüsselbach S (2000) Gene Therapy 7:224
74. de Graaf M, Pinedo HM, Oosterhoff D, van der Meulen-Muileman IH, Gerritsen WR, Haisma HJ, Boven E (2004) Human Gene Ther 15:229
75. Hay MP, Wilson WR, Denny WA (2005) Bioorg Med Chem 13:4043
76. Mauger AB, Burke PJ, Somani HH, Friedlos F, Kox RJ (1994) J Med Chem 37:3452
77. Niculescu-Duvaz I, Niculescu-Duvaz D, Friedlos F, Spooner R, Martin J, Marais R, Springer C (1999) J Med Chem 42:2485
78. Vrudhula VM, Senter PD, Fischer KJ, Wallace PM (1993) J Med Chem 36:919
79. Maeda H, Wu J, Sawa T, Matsumura Y, Hori K (2000) J Contr Release 65:271
80. Duncan R, Kopeckova-Rejmanova P, Strohalm J, Hume I, Cable HC, Pohl J, Llyod JB, Kopecek J (1987) Br J Cancer 55:165
81. Duncan R (1992) Anti-Cancer Drugs 3:175
82. Vasey PA, Kaye SB, Morrison R, Twelves C, Wilson P, Duncan R, Thomson AH, Murray LS, Hilditch TE, Murray Y, Burtles S, Fraier D, Frigerio E, Cassidy J (1999) Clin Cancer Res 5:83
83. Satchi-Fainaro R, Connors TA, Duncan R (2001) Brit J Cancer 85:1070
84. Seymour LW, Ferry DR, Anderson D, Hesslewood S, Julyan PJ, Poyner R, Doran J, Young AM, Burtles S, Kerr DJ (2002) J Clin Oncol 20:1668
85. Duncan R, Vicent MJ, Greco F, Nicholson RI (2005) Endocrine-Related Cancer 12:S189
86. Satchi-Fainaro R, Hailu A, Davies JW, Summerford C, Duncan R (2003) Bioconjugate Chem 14:797
87. Senter PD, Svensson HP, Schreiber GJ, Rodriguez JL, Vrudhula VM (1995) Bioconjugate Chem 6:389
88. Minton NP, Mauchline ML, Lemmon MJ, Brehm JK, Fox M, Michael NP, Giaccia A, Brown JM (1995) FEMS Microbiol Rev 17:357
89. Theys J, Pennington O, Dubois L, Anlezark G, Vaughan T, Mengesha A, Landuyt W, Anné J, Burke PJ, Dûrre P, Wouters BG, Minton NP, Lambin P (2006) Brit J Cancer 95:1212

90. Dang LH, Bettegowda C, Huso DL, Kinzler KW, Vogelstein B (2001) Proc Natl Acad Sci USA 98:15155

91. Agrawal N, Bettegowda C, Cheong I, Geschwind JF, Drake CG, Hipkiss EL, Tatsumi M, Dang LH, Diaz LA Jr, Pomper M, Abusedera M, Wahl RL, Kinzler KW, Zhou S, Huso DL, Vogelstein B (2004) Proc Natl Acad Sci USA 101:15172

92. Cheong I, Huang X, Bettegowda C, Diaz LA Jr, Kinzler KW, Zhou S, Vogelstein B (2006) Science 314:1308

93. Robinson MA, Charlton ST, Garnier P, Wang XT, Davis SS, Perkins AC, Frier M, Duncan R, Savage TJ, Wyatt DA, Watson SA, Davis BG (2004) Proc Natl Acad Sci USA 101:14527

94. Denny WA (2004) Cancer Investigat 22:604

95. DeFeo-Jones D, Garsky VM, Wong BK, Feng DM, Bolyar T, Haskell K, Kiefer DM, Leander K, McAvoy E, Lumma P, Wai J, Senderak ET, Motzel SL, Keenan K, Van Zwieten M, Lin JH, Freidinger R, Huff J, Oliff A, Jones RE (2000) Nat Med 6:1248

96. Denmeade SR, Nagy A, Gao J, Lilja H, Schally AV, Isaacs JT (1998) Cancer Res 58:2537

97. Levesque M, Yu H, D'Costa M, Tadross L, Diamandis EP (1995) J Clin Lab Anal 9:375

98. Catalona WJ, Smith DS, Ratliff TL, Dodds KM, Coplen DE, Yuan JJ, Petros JA, Andriole GL (1991) New Engl J Med 324:1156

99. Catalona WJ, Smith DS, Ratliff TL, Dodds KM, Coplen DE, Yuan JJ, Petros JA, Andriole GL (1991) Erratum. N Engl J Med 325:1324

100. Garsky VM, Lumma PK, Feng DM, Wai J, Ramjit HG, Sardana MK, Oliff A, Jones RE, DeFeo-Jones D, Freidinger RM (2001) J Med Chem 44:4216

101. DiPaola RS, Rinehart J, Nemunaitis J, Ebbinghaus S, Rubin R, Capanna T, Ciardella M, Doyle-Lindrud S, Goodwin S, Fontaine M, Adams N, Williams A, Schwartz M, Winchell G, Wickersham K, Deutsch P, Yao SL (2002) J Clin Oncol 20:1874

102. Wong BK, DeFeo-Jones D, Jones RE, Garsky VM, Feng DM, Oliff A, Chiba M, Ellis JD, Lin JH (2001) Drug Metab Dispos 29:313

103. Foekens JA, Peters HA, Look MP, Portengen H, Schmitt M, Kramer MD, Brünner N, Jäniccke F, Meijer-van Gelder ME, Henzen-Logmans SC, van Putten WL, Klijn JG (2000) Cancer Res 60:636

104. Noel A, Albert V, Bajou K, Bisson C, Devy L, Frankenne F, Maquoi E, Masson V, Sounni NE, Foidart JM (2001) Surg Oncol Clin N Am 10:417

105. Ganesh S, Sier CFM, Heerding MM, Vankrieken JHJM, Griffioen G, Welvaart K, van de Velde CJ, Verheijen JH, Lamers CB, Verspaget HW (1996) Cancer 77:1035

106. Ganesh S, Sier CFM, Griffioen G, Vloedgraven HJM, de Boer A, Welvaart K, van de Velde CJ, van Krieken JH, Verheijen JH, Lamers CB, Verspaget HW (1994) Cancer Res 54:4065

107. Chakravarty PK, Carl PL, Weber MJ, Katzenellenbogen JA (1983) J Med Chem 26:638

108. Eisenbrand G, Lauck-Birkel S, Tang WC (1996) Synthesis 1246

109. de Groot FMH, de Bart ACW, Verheijen JH, Scheeren HW (1999) J Med Chem 42:5277

110. Devy L, de Groot FMH, Blacher S, Hajitou A, Beusker PH, Scheeren HW, Foidart JM, Noel A (2004) FASEB 18:565

111. Egeblad M, Werb Z (2002) Nat Rev Cancer 2:161

112. Strongin A, Collier I, Bannikov G, Marmer BL, Grant GA, Goldberg GI (1995) J Biol Chem 270:5331

113. McCawley L, Matrisian L (2001) Curr Opin Cell Biol 13:534

114. Kline T, Torgov MY, Mendelsohn BA, Cerveny CG, Senter PD (2004) Mol Pharm 1:9
115. Albright CF, Graciani N, Han W, Yue E, Stein R, Lai Z, Diamond M, Dowling R, Grimminger L, Zhang SY, Behrens D et al. (2005) Mol Cancer Ther 4:751
116. Chen JM, Dando PM, Rawlings NB, Brown MA, Yong ME, Stevens RA, Hewitt E, Watts C, Barrett AJ (1997) J Biol Chem 272:8090
117. Murthy RV, Arbman G, Gao J, Roodman GD, Sun XF (2005) Clin Cancer Res 11:2293
118. Liu C, Sun C, Huang H, Janda K, Edgington T (2003) Cancer Res 63:2957
119. de Jong J, Klein I, Bast A, van der Vijgh WJ (1992) Cancer Chemother Pharmacol 31:156
120. Fernandez AM, Van Derpoorten K, Dasnois L, Lebtahi K, Dubois V, Lobl TJ, Gangwar S, Oliyai C, Lewis ER, Shochat D, Trouet A (2001) J Med Chem 44:3750
121. Trouet A, Passioukov A, Van derpoorten K, Fernandez AM, Abarca-Quinones J, Baurain R, Lobl TJ, Oliyai C, Shochat D, Dubois V (2001) Cancer Res 61:2843
122. Dubois V, Dasnois L, Lebtahi K, Collot F, Heylen N, Havaux N, Fernandez AM, Lobl TJ, Oliyai C, Nieder M, Shochat D, Yarranton GT, Trouet A (2002) Cancer Res 62:2327
123. Dubois V, Nieder M, Collot F, Negrouk A, Nguyen TT, Gangwar S, Reitz B, Wattiez R, Dasnois L, Trouet A (2006) Eur J Cancer 42:3049
124. Low PS (2004) Adv Drug Deliver Rev 56:1055
125. Reddy JA, Allagada VM, Leamon CP (2005) Curr Pharm Biotechnol 6:131
126. Hilgenbrink AR, Low PS (2005) J Pharm Sci 94:2135
127. Yoo HS, Park TG (2004) J Control Rel 96:273
128. Yoo HS, Park TG (2004) J Control Rel 100:247
129. Goren D, Horowitz AT, Tzemach D, Tarshish M, Zalipsky S, Gabizon A (2000) Clin Cancer Res 6:1949
130. Shmeeda H, Mak L, Tzemach D, Astrahan P, Tarshish M, Gabizon A (2006) Mol Cancer Ther 5:818
131. Gabizon AA, Shmeeda H, Zalipsky S (2006) Liposome Res 16:175
132. Pan XQ, Lee RJ (2004) Expert Opin Drug Deliv 1:7
133. Pan XQ, Lee RJ (2005) Anticancer Res 25(1A):343
134. Saul JM, Annapragada AV, Bellamkonda RV (2006) J Control Rel 114:277
135. Mishra PR, Jain NK (2003) Drug Deliv 10:277
136. Ren Y, Wong SM, Lim LY (2007) Bioconjugate Chem 18:836
137. Vrudhula VM, Senter PD, Fisher KJ, Wallace PM (1993) J Med Chem 36:919
138. Zhang Q, Xiang G, Zhang Y, Yang K, Fan W, Lin J, Zeng F, Wu J (2006) J Pharm Sci 95:2266
139. Nagy A, Schally AV, Halmos G, Armatis P, Cai RZ, Csernus V, Kovács M, Koppán M, Szepesházi K, Kahán Z (1998) Proc Natl Acad Sci USA 95:1794
140. Koppán M, Nagy A, Schally AV, Arencibia JM, Plonowski A, Halmos G (1998) Cancer Res 58:4132
141. Engel JB, Schally AV, Halmos G, Baker B, Nagy A, Keller G (2005) Cancer 104:1312
142. Keller G, Engel JB, Schally AV, Nagy A, Hammann B, Halmos G (2005) Int J Cancer 114:831
143. Cuttitta F, Carney DN, Mulshine J, Moody TW, Fedorko J, Fischler A, Minna JD (1985) Nature 316:823
144. Nagy A, Armatis P, Cai RZ, Szepeshazi K, Halmos G, Schally AV (1997) Proc Natl Acad Sci USA 94:652
145. Keller G, Schally AV, Nagy A, Halmos G, Baker B, Engel JB (2005) Cancer 104:2266
146. Engel JB, Schally AV, Halmos G, Baker B, Nagy AA, Keller G (2005) Endocr Relat Cancer 12:999

147. Engel JB, Schally AV, Halmos G, Baker B, Nagy AA, Keller G (2005) Endocr Relat Cancer 73:851
148. Engel JB, Schally AV, Halmos G, Baker B, Nagy AA, Keller G (2005) Eur J Cancer 41:1824
149. Stangelberger A, Schally AV, Letsch M, Szepeshazi K, Nagy A, Halmos G, Kanashiro CA, Corey E, Vessella R (2006) Int J Cancer 118:222
150. Wang X, Kebs LJ, Al-Nuri M, Pudavar H, Ghosal S, Liebow C, Nagy A, Schally AV, Prasad PN (1996) Proc Natl Acad Sci USA 96:11081
151. Bajo AM, Schally AV, Halmos G, Nagy A (2003) Clinical Cancer Research 9:3742
152. Keller G, Schally AV, Gaiser T, Nagy A, Baker B, Halmos G, Engel JB (2005) Clin Cancer Res 11:5549
153. Keller G, Schally AV, Gaiser T, Nagy A, Baker B, Halmos G, Engel JB (2005) Eur J Cancer 41:2196
154. Keller G, Schally AV, Gaiser T, Nagy A, Baker B, Westphal G, Halmos G, Engel JB (2005) Cancer Res 65:5857
155. Nagy A, Schally AV (2005) Biol Reprod 73:851
156. de Groot FMH, Broxterman HJ, Adams HP, van Vliet A, Tesser GI, Elderkamp YW, Schraa AJ, Kok RJ, Molema G, Pinedo HM, Scheeren HW (2002) Mol Cancer Ther 1:901
157. Ceh B, Winterhalter M, Frederik PM, Vallner JJ, Lasic DD (1997) Adv Drug Deliv Rev 24:165
158. Maeda H, Wu J, Sawa T, Matsumura Y, Hori K (2000) J Control Release 65:271
159. Dvorak HF (1990) Prog Clin Biol Res 354A:317
160. Yuan F, Lcunig M, Huang SK, Berk DA, Papahadjopoulos D, Jain RK (1994) Cancer Res 54:3352
161. Harvie P, Wong FM, Bally MB (2000) J Pharm Sci 89:652
162. Ruoslahti E (1996) Annu Rev Cell Dev Biol 12:697
163. Schiffelers RM, Koning GA, ten Hagen TL, Fens MH, Schraa AJ, Janssen AP, Kok RJ, Molema G, Storm G (2003) J Control Release 91:115
164. Xiong XB, Huang Y, Lu WL, Zhang X, Zhang H, Nagai T, Zhang Q (2005) J Control Release 107:262
165. Bibbya DC, Talmadgeb JE, Dalala MK, Kurzb SG, Chytilb KM, Barrya SE, Shanda DG, Steier M (2005) Int J Pharm 293:281–290
166. Rousselle C, Clair P, Lefauconnier JM, Kaczorek M, Scherrmann JM, Temsamani J (2000) Mol Pharmacol 57:679
167. Rousselle C, Smirnova M, Clair P, Lefauconnier JM, Chavanieu A, Calas B, Scherrmann JM, Temsamani J (2001) J Pharmacol Exp Ther 296:124
168. Mazel M, Clair P, Rousselle C, Vidal P, Scherrmann JM, Mathieu D, Temsamani J (2001) Anti-Cancer Drugs 12:107
169. Meyer-Losic F, Quinonero J, Dubois V, Alluis B, Dechambre M, Michel M, Cailler F, Fernandez AM, Trouet A, Kearsey J (2006) J Med Chem 49:6908

Top Curr Chem (2008) 283: 141–170
DOI 10.1007/128_2007_4
© Springer-Verlag Berlin Heidelberg
Published online: 24 November 2007

Anthracycline–Formaldehyde Conjugates and Their Targeted Prodrugs

Tad H. Koch (✉) · Benjamin L. Barthel · Brian T. Kalet · Daniel L. Rudnicki · Glen C. Post · David J. Burkhart

Department of Chemistry and Biochemistry, University of Colorado,
Boulder, CO 80309-0215, USA
Tad.koch@colorado.edu

Abstract The sequence of research leading to a proposal for anthracycline cross-linking of DNA is presented. The clinical anthracycline antitumor drugs are anthraquinones, and as such are redox active. Their redox chemistry leads to induction of oxidative stress and drug metabolites. An intermediate in reductive glycosidic cleavage is a quinone methide, once proposed as an alkylating agent of DNA. Subsequent research now implicates formaldehyde as a mediator of anthracycline-DNA cross-linking. The cross-link at 5′-GC-3′ sites consists of a covalent linkage from the amino group of the anthracycline to the 2-amino group of the G-base through a methylene from formaldehyde, hydrogen bonding from the 9-OH to the G-base on the opposing strand, and hydrophobic interactions through intercalation of the anthraquinone. The combination of these interactions has been described as a *virtual cross-link* of DNA. The origin of the formaldehyde in vivo remains a mystery. In vitro, doxorubicin reacts with formaldehyde to give firstly a monomeric oxazolidine, doxazolidine, and secondly a dimeric oxazolidine, doxoform. Doxorubicin reacts with formaldehyde in the presence of salicylamide to give the N-Mannich base conjugate, doxsaliform. Doxsaliform is several fold more active in tumor cell growth inhibition than doxorubicin, but doxazolidine and doxoform are orders of magnitude more active than doxorubicin. Exploratory research on the potential for doxsaliform and doxazolidine as targeted cytotoxins is presented. A promising lead design is pentyl PABC-Doxaz, targeted to a carboxylesterase enzyme overexpressed in liver cancer cells and/or colon cancer cells.

Keywords Androgen receptor · $\alpha_V\beta_3$ integrin · Carboxylesterase · Doxazolidine · Doxsaliform · Estrogen receptor · Pentyl PABC-Doxaz

Abbreviations

AR	androgen receptor
C	L-cysteine or deoxycytidine
CES	carboxylesterase
D	L-aspartic acid
Dox	doxorubicin
Doxaz	doxazolidine
DoxF	doxoform
DoxSF	doxsaliform
DTT	dithiothreitol
ER	estrogen receptor
f	D-phenylalanine
G	glycine or deoxyguanidine
(GC)$_4$	5′-GCGCGCGC-3′
GFP	green fluorescent protein
IC$_{50}$	concentration of drug that inhibits half the growth or binding
K	L-lysine
MDR	multidrug resistance
NCI	National Cancer Institute,
NSCL	non small cell lung
PABC	p-aminobenzylcarbamate
R	L-arginine
SCL	small cell lung
T	L-threonine
Tam	tamoxifen
Teg	triethyleneglycol
V	L-valine
W	L-tryptophan
Y	L-tyrosine

1
Introduction

In this section of Chap. 3, the focus is the chemistry of the clinical anthracyclines, doxorubicin, epidoxorubicin, daunorubicin, and idarubicin, that leads to alklyation of DNA as part of a cytotoxic mechanism (see Fig. 1 for structures and numbering system). For the clinical anthracyclines, crosslinking is mediated by formaldehyde, and natural and synthetic anthracycline-aldehyde conjugates at the 3′-amino group show enhanced cytotoxic activity, especially to resistant cancer cells. This discovery has led to the design and synthesis of prodrugs of aldehyde conjugates targeted to receptors and/or enzymes overexpressed by cancer cells as improved therapeutics. Targeted prodrugs of anthracycline-aldehyde conjugates is a second focus of this chapter. Phillips

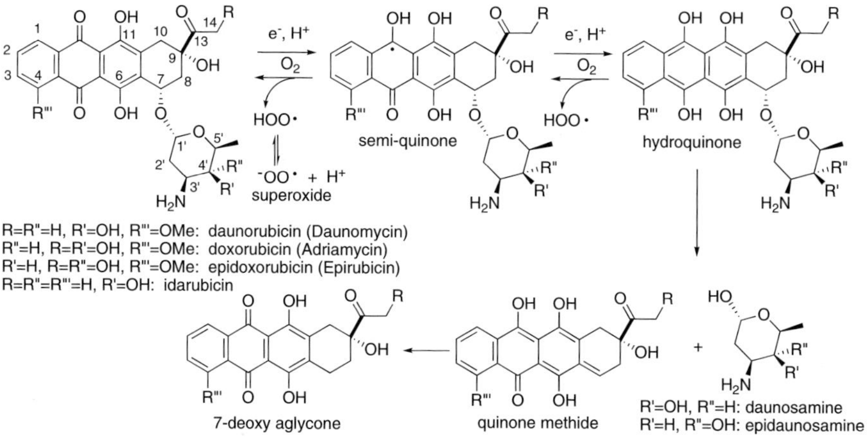
R=R"=H, R'=OH, R'"=OMe: daunorubicin (Daunomycin)
R"=H, R=R'=OH, R'"=OMe: doxorubicin (Adriamycin)
R'=H, R=R"=OH, R'"=OMe: epidoxorubicin (Epirubicin)
R=R"=R'"=H, R'=OH: idarubicin
semi-quinone
hydroquinone
superoxide
7-deoxy aglycone
quinone methide
R'=OH, R"=H: daunosamine
R'=H, R"=OH: epidaunosamine

Fig. 1 Structures of clinically important anthracycline anti-tumor drugs with numbering system and redox chemistry that leads to oxidative stress and reductive glycosidic cleavage via the quinone methide

and coworkers have published an excellent review of the history of the discovery of formaldehyde-mediated anthracycline DNA alkylation from a complementary perspective [1].

Early on, the anthracyclines were observed to be good intercalators in DNA, and drug DNA intercalation was proposed to be involved in the cytotoxic mechanism [2–7]. Later experiments implicated that anthracycline induction of DNA double strand breaks through the effect of intercalation on the processing of DNA by topoisomerase II [8–10]. Circumstantial evidence for anthracycline alkylation of DNA as part of the cytotoxic mechanism also appeared in the literature [11–13], but establishing the structure of the drug-DNA cross-link proved to be a challenge, primarily because of its instability. New results now indicate that the cross-links induce cell death by a topoisomerase II independent mechanism [14].

Identification of anthracycline metabolites, especially the product of reduction at the quinone, the 7-deoxyaglycones (Fig. 1), suggested that redox chemistry occurs in vivo [15]. Of early interest was the quinone methide intermediate from reductive glycosidic cleavage (Fig. 1) that could potentially serve as an alkylating agent at nucleophilic sites in DNA [16, 17]. In this context, the anthracyclines were classified as bioreductively activated natural products [18]. Extensive research by us and by others, however, failed to show the quinone methide alklyation of DNA, although, we did observe covalent bond formation between the amino group of a G-base and the quinone methide from reductive cleavage of the semi-synthetic anthracycline Menogaril (Fig. 2) [19]. The quinone methide from the reductive cleavage of daunorubicin proved to be more nucleophilic than electrophilic as exemplified most simply by reaction with a proton to give the 7-deoxyaglycone [16].

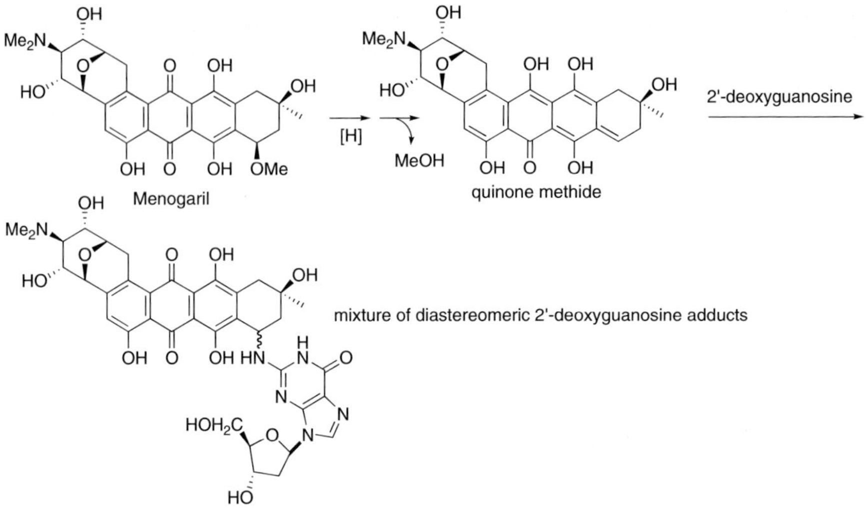

Fig. 2 Reductive glycosidic cleavage of Menogaril and the reaction of its quinone methide with 2'-deoxyguanosine

Even though anthracycline redox chemistry does not lead directly to DNA alklyation, it is important for the induction of oxidative stress, yet another cytotoxic mechanism [17, 20–23]. Under aerobic conditions, reduction of the quinone functional group to semiquinone or hydroquinone redox states followed by back oxidation by dioxygen gives superoxide in equilibrium with hydroperoxy radical that can disproportionate to hydrogen peroxide. This is then a catalytic process for the production of elements of oxidative stress and is thought to be important in the treatment-limiting side effect of anthracycline therapy, cardiotoxicity [24]. Cardiotoxicity in part results from anthracycline-accumulation in heart cells through attraction as a cation to the abundant, negatively charged, mitochrondial membrane lipid, cardiolipin, and low levels of enzymes in heart cells that neutralize oxidative stress [25, 26]. Minotti and coworkers [27] have recently explored this aspect of the redox chemistry in detail, especially with respect to the role of iron that accentuates the oxidative stress, as reported in their section of this volume of Topics in Current Chemistry. Kalyanaraman and coworkers have also presented interesting results that point to the doxorubicin induction of oxidative stress being more important for triggering apoptosis in normal cells, cardiomyocytes and endothelial cells, than in tumor cells [28]. Further, doxorubicin activation of the transcription factor p53 appears to be more important for the induction of tumor cell apoptosis than normal cell apoptosis.

In 1994, Philips and coworkers reported substantial evidence for doxorubicin cross-linking of DNA [29]. The reaction conditions were doxorubicin, dithiothreitol (DTT), FeCl$_3$, and various forms of DNA in transcription buffer. The assay for DNA cross-linking was the observation of transcription blockages. Two types of unstable transcription blockages were observed, one with a 4 h half-life and one with a 30 h half-life. The shorter-lived blockages were proposed to be drug adducts at isolated G-bases, and the longer lived blockages were proposed to be double strand cross-links at a 5'-GC-3' sites. The quinone methide transient was proposed as a possible reactive intermediate for forming a bond to the amino group of a G-base analogous to that observed earlier with the quinone methide from reductive cleavage of Menogaril [19]. For the cross-link, the second bond was proposed to occur by reaction at the ketone side chain at the 9-position. These and other mechanisms of anthracycline cytotoxicity were critically reviewed by Gewirtz in 1999 [30].

2
Virtual Cross-Linking of DNA by Anthracyclines

Upon learning the doxorubicin-DNA cross-linking results from the Phillips lab, our research group decided to study the molecular nature of the cross-link using negative ion electrospray mass spectrometry. For our studies of cross-linking with daunorubicin and doxorubicin under the Phillips' redox conditions, we used 5'-GCGCGCGC-3' ($(GC)_4$) as a simple self-complementary oligonucleotide with a reasonable melting temperature bearing 5'-GC-3' sites. Multiple products with both DNA and anthracycline chromophores were evident by HPLC [31, 32]. The major product showed two anthracycline chromophores per duplex DNA as determined by optical density. Negative ion electrospray mass spectrometry showed a molecular ion with mass equal to duplex DNA plus two anthracyclines plus two extra carbons. Several years earlier, Wang and coworkers reported a crystal structure for duplex 5'-CGCGCG-3' covalently bound to two daunorubicin molecules [33]. The bonding was from the 3'-amino group of each daunorubicin to the 2-amino group of a G-base via a methylene originating from formaldehyde as an impurity in the crystallization solvent. This crystal structure together with the molecular mass of our major product suggested an analogous structure with just two extra base pairs, one at each end of the double-stranded DNA. Clearly, our choice of oligonucleotide was serendipitous, and identification of the product structure benefited immensely from the Wang crystal structure. The Wang crystal structure also showed hydrogen bonding from the 9-OH group of the daunorubicin to the amino group and nitrogen at the 1-position of the G-base on the opposing strand [33, 34]. We proposed the term *virtual cross-link* for the total bonding interaction between DNA and daunorubicin consisting of the covalent bonding to one strand via formaldehyde, hydrogen bonding to the other strand, and hydrophobic interactions with both strands [32]. The structure with two symmetrically arranged *virtual cross-links* is shown schematically in Fig. 3 with 5'-GCGCGCGC-3' and in three dimensions in Fig. 3 with 5'-CGCGCG-3' based on the Wang crystal structure. Subsequently, the

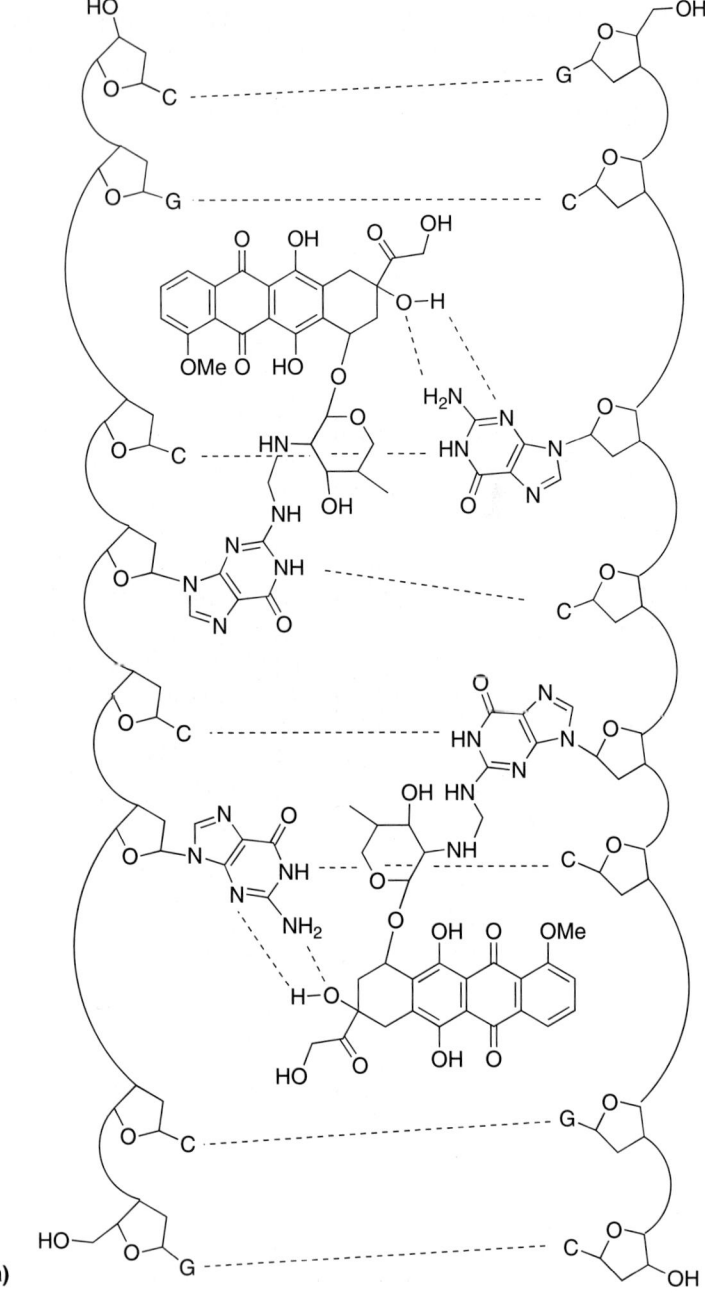

Fig. 3 a Schematic structure of double stranded 5′-GCGCGCGC-3′ DNA with two doxo-rubicin-formaldehyde *virtual cross-links* at 5′-GC-3′ sites. *Virtual cross-linking* of DNA occurs through covalent bonding to a G-base on one strand and hydrogen bonding to a G-base on the opposing strand

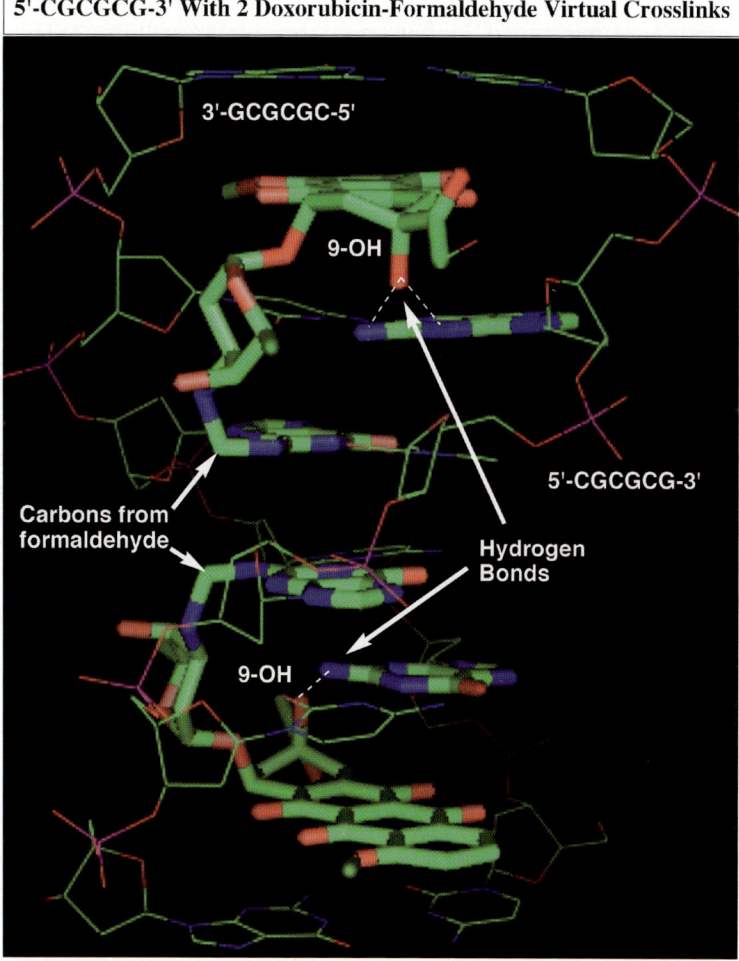

5'-CGCGCG-3' With 2 Doxorubicin-Formaldehyde Virtual Crosslinks

Fig. 3 b Three dimensional structure of double stranded 5'-CGCGCG-3' with two doxo-rubicin-formaldehyde *virtual crosslinks*. DNA is shown as a line drawing and the *virtual cross-links* as stick drawings. The picture was created from the coordinates of the Wang crystal structure [33] (Rutgers Protein Data Bank, I D number 1D33) by adding the 14-hydroxyl in Chem-3D and displaying the result in PyMol

virtual cross-link with doxorubicin in DNA was established by large molecule NMR spectroscopy [35] and with epidoxorubicin in DNA by crystallogra-phy [36]. NMR was also used to estimate the effect of the *virtual cross-link* on the kinetic stability of double stranded DNA by measuring the rate of strand exchange with DNA bearing one *virtual cross-link* or one intercalated doxoru-bicin, or with unintercalated DNA. One intercalated doxorubicin stabilizes the duplex 3.9-fold, but one *virtual cross-link* stabilizes the duplex 637-fold [35]. Experiments comparing the denaturation of DNA with *virtually cross-linked*

DNA indicate that a *virtual cross-link* is the equivalent of four additional GC base pairs or six additional AT base pairs on duplex stability [35]. *Virtual cross-linking* has also been demonstrated with natural DNA sequences [37].

What was the source of the formaldehyde? The redox conditions were those of the Fenton reaction to produce superoxide, hydrogen peroxide, and hydroxyl radical through sequential one electron reductions of dioxygen mediated by iron. DTT is not a natural thiol but the natural thiol glutathione is a functional equivalent. DTT can also be replaced with the enzymatic system xanthine oxidase/NADH [29, 38]. Transcription buffer contains Tris, trishydroxymethyl-aminomethane, which like many other organic buffers is oxidized under Fenton conditions to formaldehyde, amongst other products [39]. The transcription buffer plus $FeCl_3$ and DTT produces large quantities of formaldehyde even in the absence of anthracycline [38]. Hence, the redox chemistry to produce the formaldehyde was independent of the anthracycline. Doxorubicin is also a substrate for Baeyer-Villiger oxidation with hydrogen peroxide at the 13-position to produce formaldehyde at ambient temperature in pH 7.4 phosphate buffer [31]. The requisite hydrogen peroxide might logically come from anthracycline redox cycling as shown in Fig. 1. Therefore, doxorubicin but not daunorubicin can function as a sacrificial source of formaldehyde. The byproducts of Baeyer-Villiger oxidation, shown in Fig. 4, are inactive.

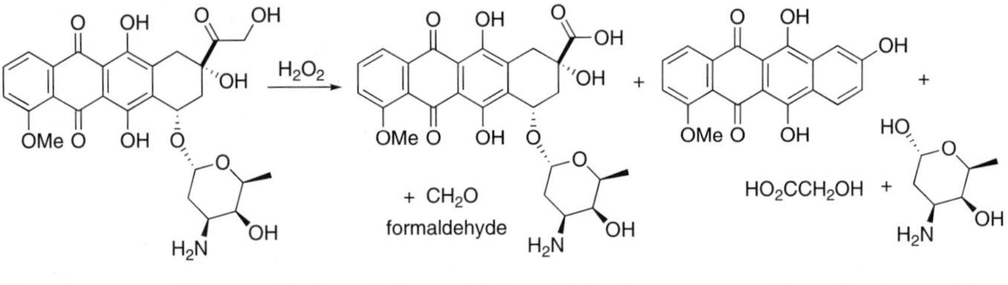

Fig. 4 Baeyer-Villiger oxidation of doxorubicin with hydrogen peroxide at the 13-position to give formaldehyde, amongst other products

What is the origin of the formaldehyde inside cells? Formaldehyde was reported to be a product of Fenton oxidation of unsaturated fatty acids of cell membrane lipids [40]. However, the experiments were performed in the Tris buffer. Our reinvestigation of this result showed that the primary source of the reported formaldehyde was oxidation of the Tris buffer and not oxidation of the unsaturated fatty acid [41]. Natural polyamines that react under Fenton conditions to release small amounts of formaldehyde are spermine and spermidine [38]. Spermine is an interesting possibility because it is associated in the major groove of DNA as a polycation for partial charge neutralization of the sugar phosphate backbone. Available iron or copper ions are required for the Fenton oxidation, and Minotti has provided a nice explanation for how anthracyclines disrupt iron homeostasis to release iron from iron storage protein [27]. Possibly, released iron first complexes with doxorubicin [42] and then catalyzes formation of formaldehyde via Fenton chemistry.

Several measurements also support higher levels of formaldehyde in cancer cells and cancer patients. Higher levels have been reported in more lymphocytic leukemia cells than normal lymphocytes [43]. Treatment of cancer cells with 0.5 μM daunorubicin for 24 h caused an order of magnitude elevation of formaldehyde concentration in MCF-7 breast cancer cells as measured by mass spectrometry of cell lysates [44, 45]. The highly resistant variant, MCF-7/Adr cells, showed only background levels of formaldehyde upon similar treatment. MCF-7/Adr cells are more than three orders of magnitude resistant to doxorubicin at least in part because they overexpress the P-170 glycoprotein drug efflux pump. Elevated formaldehyde has been detected in the urine of doxorubicin-treated rats [46], in the urine of patients with bladder and prostate cancer [47], and in the breath of tumor-bearing mice and cancer patients [48].

Two possible general mechanisms were proposed by us and by others for the *virtual cross-linking* of DNA by the combination of daunorubicin, or doxorubicin and formaldehyde [32, 37]. One has the anthracycline reacting first with formaldehyde and the conjugate reacting with DNA, and the other has DNA reacting first with the formaldehyde and its conjugate reacting with the anthracycline. The results of experiments to be described in Sects. 3, 4 and 5 point to the former mechanism as more likely.

3
Daunoform, Doxoform, Epidoxoform, and Related Alkylating Anthracyclines

Virtual cross-linking of DNA by the combination of daunorubicin, or doxorubicin and formaldehyde, prompted the synthesis of anthracycline-formaldehyde conjugates as improved therapeutics. A two- phase reaction in a chloroform/pH 6 buffer of daunorubicin or doxorubicin with formalin, a water/methanol solution of formaldehyde, yielded in the chloroform phase a conjugate with two molecules of anthracycline as oxazolidines bonded together with a third molecule of formaldehyde as shown in Fig. 5 [49]. The compounds were initially characterized spectroscopically and assigned the common names daunoform and doxoform. Doxoform was subsequently characterized in the solid state by crystallography in which it exhibits a compact structure with the daunosamine sugar rings in a twist boat conformation, as shown also in Fig. 5 [50]. The reaction of doxorubicin free base with solid paraformaldehyde in chloroform yielded first the monomeric species doxazolidine and then doxoform. The reaction can be terminated at the monomeric stage as indicated by NMR through removal of the excess paraformaldehyde by filtration [50]. In solution, the daunosamine sugars of doxorubicin, doxa-

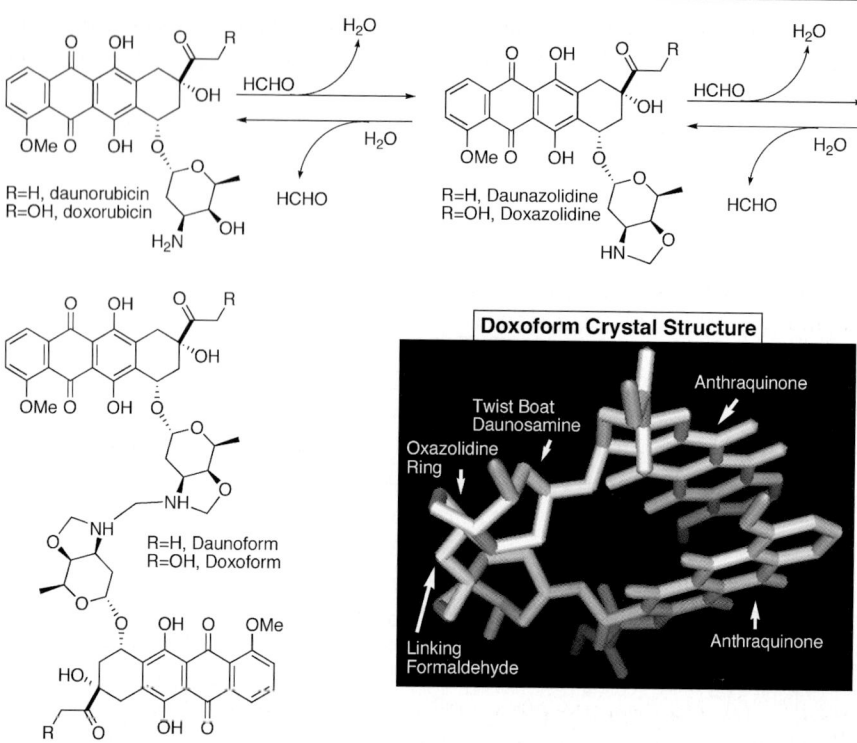

Fig. 5 Reaction of daunorubicin and doxorubicin with formaldehyde to give first dauna-zolidine and doxazolidine, respectively, and subsequently daunoform and doxoform. The three-dimensional picture of doxoform was created in PyMol using coordinates from the crystal structure [50]

zolidine, and doxoform all appear to exist in chair conformations as indicated by NMR coupling constants for protons at the 1' and 2'-positions.

Daunoform and doxoform are both hydrolytically unstable with respect to the loss of formaldehyde and the reformation of the clinical drugs. Intermediates in the hydrolyses are the monomeric species daunazolidine and doxazolidine, respectively. The hydrolysis of doxoform has been studied more extensively than the hydrolysis of daunoform. Measurements of the growth inhibition of MCF-7/Adr cells provide estimates of the half-lives of doxoform and doxazolidine, with respect to complete hydrolysis to doxorubicin. These cells are highly resistant to doxorubicin but not to doxoform or doxazolidine. Hence, measurement of growth inhibition as a function of time for hydrolysis prior to inoculation of cells gives a measure of half-lives for doxoform and doxazolidine of 1.5 min and 1 min, respectively, in cell growth media at 37 °C. The half-life of doxoform is extended to 3 min in human serum [50] and to 10 min in bovine serum [51]. A rationale for a longer half-life in 100% bovine serum has not been completely established but appears to result from the association with bovine serum albumin that makes up 50% of the serum proteins. The 10% fetal bovine serum in cell growth media has almost no effect on stability [51].

In spite of its short half-life, doxoform shows superior tumor cell growth inhibition relative to doxorubicin with one to four orders of magnitude lower IC_{50} values. The IC_{50} is the concentration that inhibits half the growth. Representative values from our lab are reported for doxoform in Table 1 compared with values for similar treatment with doxorubicin. In the few cases for which measurements were made, doxazolidine shows the same IC_{50} values as does doxoform. This is not surprising because doxoform is just a rapid prodrug of doxazolidine. The more dramatic differences between doxorubicin and doxoform occur with the more resistant cell lines. For example, a four order of magnitude difference occurs with the highly resistant MCF-7/Adr breast cancer cell line and a three order of magnitude difference occurs with the resistant SHP77 small cell lung cancer cell line. MCF-7/Adr cells respond to doxazolidine in spite of being p53 negative and MDR positive.

A treatment-limiting, chronic side effect of doxorubicin therapy is cardiotoxicity. As discussed earlier, this side effect results at least in part from the accumulation of positively charged doxorubicin in cardiomyocytes because of the abundance of mitochondria that bear the negatively charged phospholipids, cardiolipin, and the susceptibility of cardiomyocytes to oxidative stress [25, 26]. As shown in Table 1, doxorubicin, doxoform, and doxazolidine all inhibit the growth of rat cardiomyocytes equally [52]. More significantly, doxorubicin inhibits the growth of cardiomyocytes better than it inhibits

Table 1 Comparison of growth inhibition of cancer cells and rat cardiomyocytes by doxorubicin, doxoform, and doxazolidine with 3 h drug treatment [49, 50, 52]. Units for IC_{50} values are nM equiv to correct for DoxF having two active compounds per molecule

Cell Line/ Compound	IC_{50} (nM equiv)				
	MCF-7 breast	MCF-7/Adr breast	SK-Hep-1 liver	Hep G2 liver	DU-145 prostate
Dox	200	10 000	100	200	200
DoxF	2	1	4	10	3
Doxaz	3	3	–	–	4
Cell Line/ Compound	MiaPaCa-2 pancreas	BxPC3 pancreas	SHP77 SCL[a]	H2122 NSCL[b]	H9c2(2-1) Rat cardio-myocytes
Dox	300	300	>1000	200	30
DoxF	3	10	2	4	30
Doxaz	–	–	–	–	30

[a] SCL, small cell lung
[b] NSCL, Non-small cell lung

the growth of all the cancer cell lines in Table 1 except the highly resistant MCF-7/Adr cell line, and doxoform inhibits the growth of all the cancer cell lines in Table 1 better than it inhibits the growth of cardiomyocytes. Because doxoform has a half-life during cell culture experiments of only 3 min [51], most of the doxoform had hydrolyzed to doxorubicin after 15 min. Hence, during treatment with doxoform, the cardiomyocytes were exposed to doxorubicin for the majority of the 3 h treatment period, and consequently, most of the growth inhibition of cardiomyocytes with doxoform probably stemmed from the resulting doxorubicin. An additional contributing factor to the lower relative cardiotoxicity of doxoform and doxazolidine may be no positive charge at physiological pH and consequently, no Coulombic attraction to cardiolipin. Although both doxoform and doxazolidine have amino nitrogens, they are covalently bonded through the formaldehyde carbon to the 4'-oxygen which significantly lowers the pKa values for their respective protonated nitrogens [53].

Doxoform has also been evaluated in the 60 human cancer cell screen of the National Cancer Institute (NCI). NCI uses a 48 h drug treatment period with growth measurement at 48 h. Again, because of the short half-life of doxoform, the cells are exposed to doxorubicin for most of the treatment period. In spite of this limitation, doxoform showed more than a log lower average IC_{50} value than doxorubicin. This was from a comparison of the average of two measurements with doxoform with the average of 1837 measurements with doxorubicin.

Concurrently, we also studied the reaction of formaldehyde with epidoxorubicin, the 4'-epimer of doxorubicin. Epidoxorubicin is of clinical interest because of lower cardiotoxicity resulting from faster clearance through the kidneys as a glucuronic acid conjugate [54–57]. Epidoxorubicin has a transvicinal aminol functionality at the sugar ring, and, consequently, forms not a dimeric oxazolidine but a dimeric molecule with two molecules of epidoxorubicin bonded together with three molecules of formaldehyde in a diazadioxabicyclic ring structure, as shown in Fig. 6 [58]. Epidoxoform hydrolyzes to monomeric acyclic formaldehyde conjugates at pH 7.4. Surprisingly, at 25 μM initial epidoxoform concentration, the hydrolysis reaction comes to an equilibrium with monomeric formaldehyde conjugates, also shown in Fig. 6. The initial ring opening of the one-carbon bridge between the two nitrogens occurs rapidly and reversibly and then the equilibrium mixture of the two dimeric species proceeds slowly to an equilibrium with the monomeric species with a half-life of 2 h.

Epidoxorubicin plus formaldehyde or epidoxoform *virtually cross-link* (GC)₄ analogous to doxorubicin plus formaldehyde or doxoform as determined by mass spectrometry and crystallography [36, 58]. Epidoxoform is less active in tumor cell growth inhibition than doxoform, although it is still more active than epidoxorubicin or doxorubicin, especially with resistant MCF-7/Adr cells [58]. A possible explanation for the difference in cytotoxi-

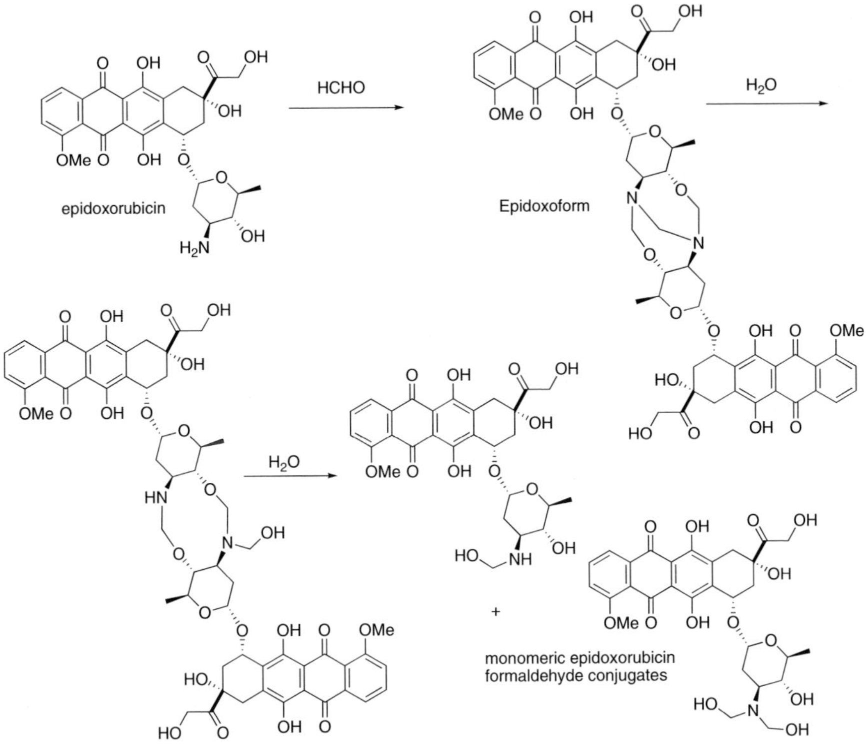

Fig. 6 Reaction of epidoxorubicin with formaldehyde to give epidoxoform, and subsequently, hydrolysis of epidoxoform to give epidoxorubicin-formaldehyde acyclic conjugates

city comes from a measurement of the hydrolytic stability of the respective *virtual cross-links*. The measurement was performed with MCF-7/Adr cells measuring the disappearance of chromophore in drug treated cells, assuming that drugs released from DNA would be rapidly pumped out of the cell by overexpressed P-170 glycoprotein. This measurement showed a half-life for loss of the drug chromophore from doxoform treated cells of 29 h and from epidoxoform treated cells of 13 h [59].

Because of the increased aqueous stability of epidoxorubicin-formaldehyde, monomeric conjugates relative to doxazolidine, epidoxoform was also investigated in vivo versus epidoxorubicin using a murine model of breast cancer. Epidoxoform is not very water soluble and so was formulated in Cremaphor/DMSO for the in vivo experiments. The maximum tolerated dose was determined to be 10 mg/kg for epidoxoform and 7.5 mg/kg for epidoxorubicin. Mice inoculated with 10^6 Gollin-B mouse mammary tumor cells were treated with the maximum tolerated dose two days following tumor cell injection and repeated in seven days. All ten mice in each group survived; however, the median tumor volume in the epidoxoform group was 68 mm^3

compared with 1764 mm^3 in the epidoxorubicin group [60]. Further, in the epidoxoform group, four mice showed no tumor; all mice in the epidoxorubicin group developed tumors.

Several natural and synthetic, alkylating anthracyclines have been investigated by others during the past two decades. These include cyanomorpholinodoxorubicin [61], barminomycin [62, 63], 2-pyrrolinodoxorubicin [64], N-5,5-diacetoxypentyl)doxorubicin [65], PNU-159682, and trimethylenedaunazolidine [66] (Fig. 7). PNU-159682 is an oxidative metabolite of

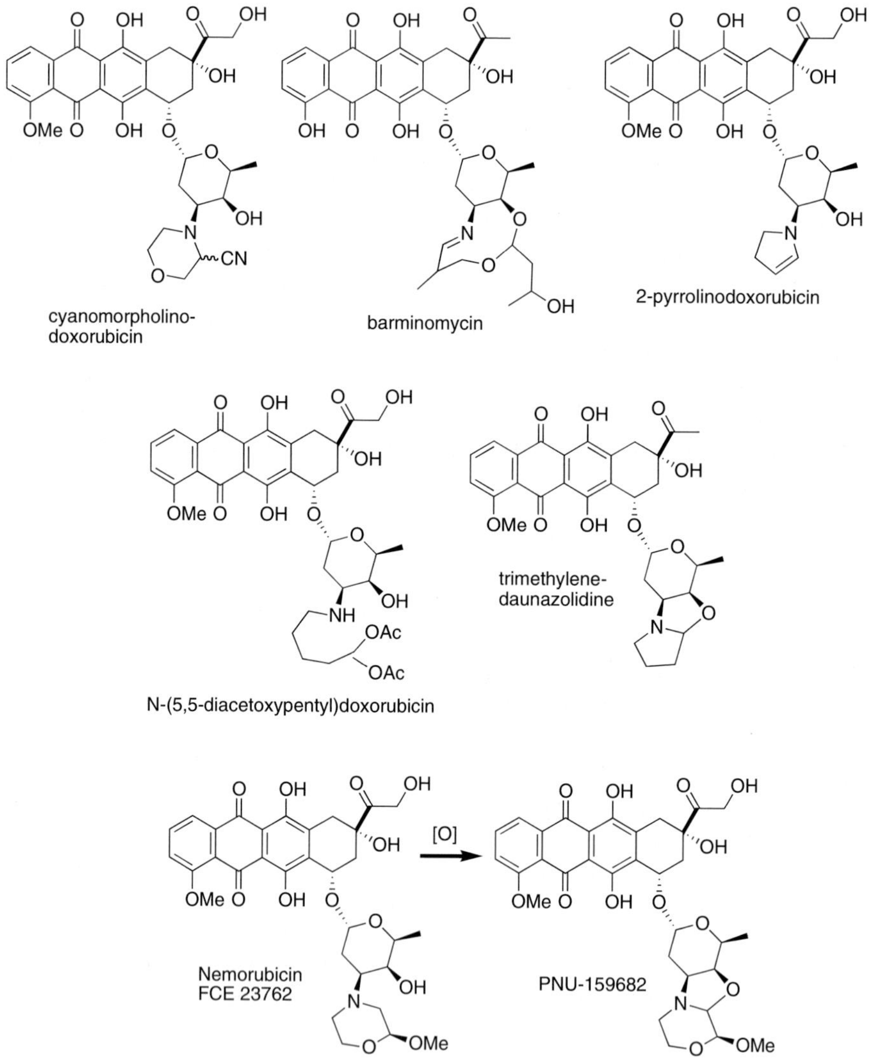

Fig. 7 Structures of other natural and synthetic, alkylating anthracyclines

Nemorubicin (FCE 23762) [67]. For a complete review of Nemorubicin, see the section by Broggini in Chap. 3. All of these examples are masked aldehydes with a second attachment of the aldehyde via a tether to the daunosamine, and some have been shown to covalently bond to G-bases at 5′-GC-3′ sites in DNA [68, 69].

4
Doxsaliform and Targeted Doxsaliform

Encouraged by the in vivo activity of epidoxoform and the superior in vitro activity of doxoform, we next explored acyclic doxorubicin formaldehyde conjugates using the formaldehyde N-Mannich base construct. Reaction of doxorubicin with formaldehyde in the presence of simple alkyl or aryl amides in DMF solvent yielded doxorubicin N-Mannich bases [70]. These proved to be too robust with respect to release of the acyclic-doxorubicin formaldehyde conjugate. Work by Bundgaard and coworkers [71] and by Loudon and coworkers [72] indicated that N-Mannich bases synthesized from salicylamide release amine-formaldehyde conjugates more rapidly through internal catalysis by the phenolic OH group. Therefore, the reaction of doxorubicin with formaldehyde in the presence of salicylamide gave doxsaliform [73]. Doxsaliform released acyclic doxorubicin-formaldehyde conjugate with a half-life of 1 h at 37 °C in pH 7.4 buffer (Fig. 8). Hence, the salicylamide N-Mannich base construct serves as a time-release trigger for delivery of acyclic doxorubicin-formaldehyde conjugate. IC_{50} values for growth inhibition of MCF-7 and MCF-7/Adr cells by doxsaliform were 4-fold and 10-fold smaller, respectively, than for growth inhibition by doxorubicin. Therefore, doxsaliform was an improvement but still lagged in activity relative to doxoform. Doxsaliform, however, did have one advantage over doxoform or epidoxoform; namely, it provided a site for attachment of a releasable targeting group that might direct the construct to a receptor overexpressed by tumor cells. Four targets have been explored using the doxsaliform strategy: an estrogen receptor and an anti-estrogen binding site for breast cancer, an androgen receptor for prostate cancer, and an $\alpha_V\beta_3$ integrin for metastatic cancer.

Targeting estrogen receptor (ER) and antiestrogen binding site (AEBS) for breast cancer. Many breast cancer cells overexpress the nuclear hormone receptor ER (more precisely ERα), and estrogen bound to ER stimulates transcription of growth factors. Tamoxifen and related non-steroidal anti-estrogens are commonly used for treatment and prevention of hormone responsive breast cancer because they or their metabolite bind to ER in competition with estrogen, and the anti-estrogen-ER complex does not stimulate transcription. On this basis ER was chosen as a receptor for targeting a doxorubicin-formaldehyde conjugate to breast cancer cells, and hydroxytamoxifen, the active metabolite of tamoxifen, was chosen as the targeting

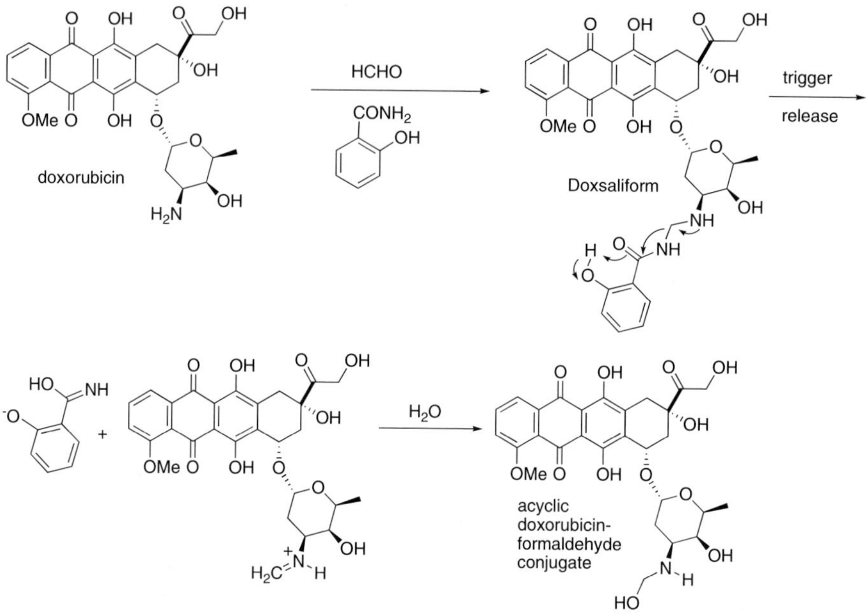

Fig. 8 Synthesis of the salicylamide N-Mannich base of doxorubicin, doxsaliform (DoxSF), and its hydrolysis to an acyclic doxorubicin-formaldehyde conjugate

group. A second receptor for non-steroidal anti-estrogens is an anti-estrogen binding site (AEBS) that is cytosolic and consists of at least four associated proteins, some of which are involved with lipid metabolism [74]. The site for attachment of the tether to hydroxytamoxifen was selected based upon the co-crystal structure of hydroxytamoxifen bound to the ligand binding domain of ERα that shows the exposure of one of the methyl groups [75]. Hydroxy-tamoxifen was thus tethered from one of the equivalent methyl groups to the salicylamide of doxsaliform [76]. Coupling to the salicylamide was achieved with an oxime functional group as shown in Fig. 9. Three ethylene glycol type tethers were explored for binding to the receptors and tumor cell growth inhibition. The best of the three with respect to receptor affinity and tumor cell growth inhibition was a triethylene glycol tether. A space-filling model of the construct with the triethylene glycol tether, DoxSF-Teg-Tam, bound to the ligand binding domain of ERα is shown in Fig. 10.

The performance of DoxSF-Teg-Tam was compared with the performance of hydroxytamoxifen in terms of binding to receptors and with the perform-ance of DoxSF and Dox in terms of tumor cell growth inhibition. The binding affinity of DoxSF-Teg-Tam to the ER from cell lysis relative to hydroxyta-moxifen was 2.5%. Hence, DoxSF-tether when bound to the methyl group of hydroxytamoxifen significantly attenuated but did not eliminate the binding to ER. The binding affinity of DoxSF-Teg-Tam to AEBS relative to hydroxyta-moxifen was 60%. The inhibition of growth of breast cancer cells is reported

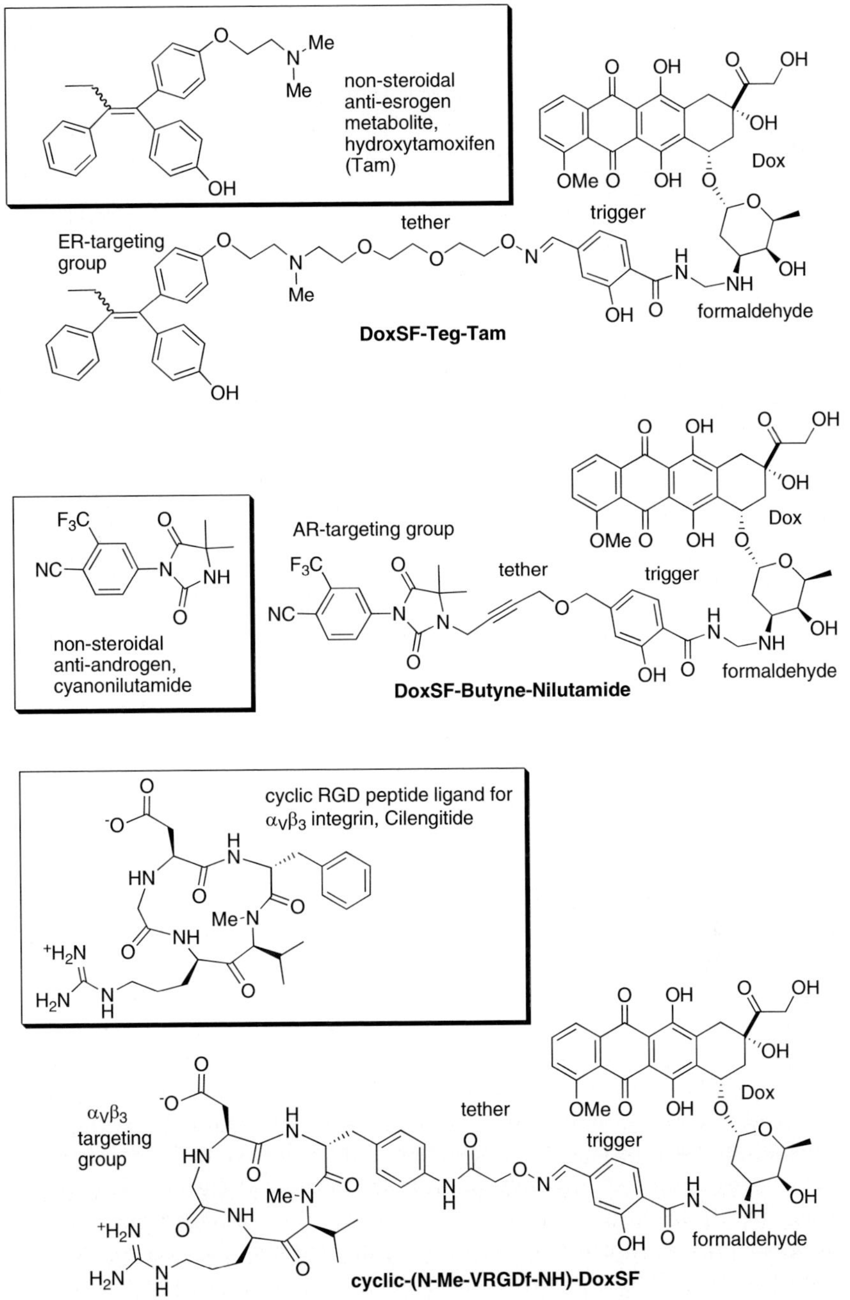

Fig. 9 Structures of targeting groups and targeted DoxSF molecules: DoxSF-Teg-Tam targeted with hydroxytamoxifen to the estrogen receptor, DoxSF-butyne-nilutamide targeted with cyanonilutamide to the androgen receptor, and cyclic-(N-Me-VRGDf-NH)-DoxSF targeted to the integrin $\alpha_V\beta_3$

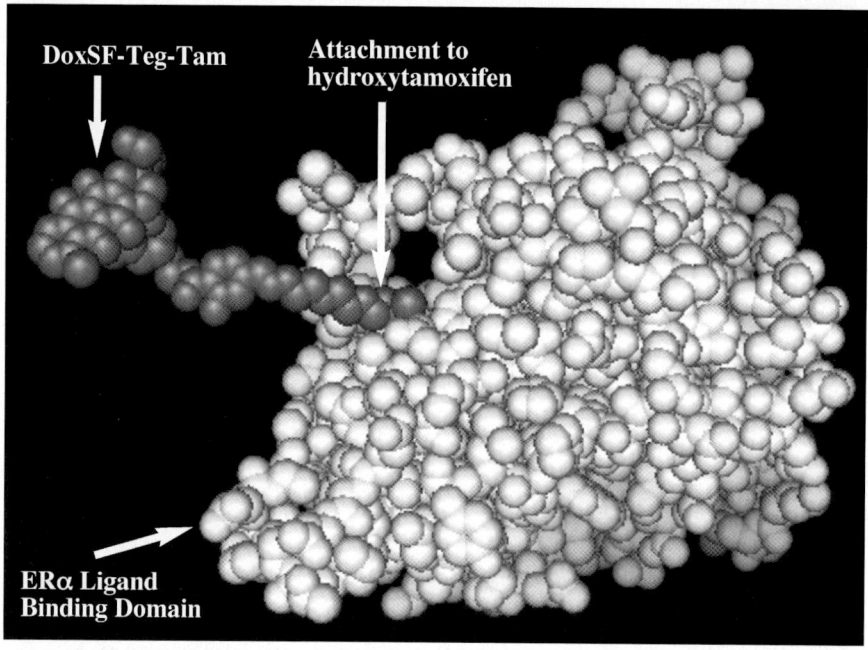

Fig. 10 A model of DoxSF-Teg-Tam bound to the ligand binding domain of the estrogen receptor (ERα). The model was created by docking DoxSF-Teg created in Chem-3D by inspection to the hydroxytamoxifen in the ERα co-crystal structure (from Shiau and co-workers [75], PDB code: 3ERT) and displayed with PyMol

in Table 2 as a function of the expression of ER, AEBS, and the multidrug resistance phenotype (MDR). The most dramatic effect on tumor cell growth inhibition was observed with the MCF-7/Adr cells that are ER negative, AEBS positive, and MDR positive. The IC$_{50}$ for growth inhibition with DoxSF-Teg-Tam was 30-fold lower than with DoxSF and 170-fold lower than with Dox. A 1 : 1 combination of DoxSF and hydroxytamoxifen showed no synergy relative to DoxSF; although, a 1 : 1 combination of Dox and hydroxytamoxifen did show synergy, independent of ER and MDR expression [76]. A combination of flow cytometry and fluorescence microscopy experiments with the various cell lines in Table 2 are consistent with the following sequence of events: 1) the targeted conjugate passively diffuses across the cytoplasmic and nuclear membranes; 2) the targeting group binds to cytosolic AEBS and/or nuclear ER; 3) the binding proteins serve to sequester the conjugate, reducing drug efflux by p-glycoprotein drug efflux pump expressed as part of the MDR phenotype; and 4) the salicylamide trigger fires, releasing the doxorubicin formaldehyde conjugate that intercalates and alkylates DNA to induce cell death [76, 77]. In contrast, Dox and DoxSF are substrates for p-glycoprotein overexpressed in MCF-7/Adr cells and are pumped out in competition with passive diffusion into the cells.

Table 2 Comparison of growth inhibition for breast cancer cells as a function of targeting, formaldehyde, and the expression of ER, AEBS, and MDR [77]

Compound/Cell line	ER/AEBS/MDR[a]	IC$_{50}$ (nM) Dox	DoxSF	DoxSF-Teg-Tam
MCF-7	+/+/−	200	70	30
MCF-7/Adr	−/+/+	10 000	2000	60
Rtx-6	+/−/−	200	60	70
MDA-MB-231	−/+/−	300	80	30
MDA-MB-435	−/+/−	150	50	40

[a] ER, estrogen receptor; AEBS, antiestrogen binding site; MDR, multidrug resistance phenotype

Targeting the androgen receptor (AR) for prostate cancer. In parallel to breast cancer cells, prostate cancer cells commonly overexpress AR, and non-steroidal anti-androgens are used for the treatment of hormone responsive prostate cancer. For targeting the AR, the non-steroidal anti-androgen, cyanonilutamide, was selected as the targeting group. Similar to hydroxytamoxifen bound to the ER, non-steroidal anti-androgens bound to the AR show limited opportunity for attachment of the tether. The point of attachment to the nilutamide was selected as the substitutable nitrogen based upon crystallography data and molecular modeling. Various tether designs were explored, all with tethering from the substitutable nitrogen of the nilutamide to the salicylamide as shown for the lead design in Fig. 9. The best tether, as determined from binding measurements to the androgen receptor, incorporated 2-butyne which gave an IC$_{50}$ for binding to the androgen receptor of 90 nM [78]. For comparison, the non-steroidal anti-androgen drugs, nilutamide and flutamide, gave binding IC$_{50}$ values of 9 and 154 nM, respectively. The 2-butyne was more successful than others probably because it maintained separation of the targeting group from the cytotoxin while occupying minimal space. In retrospect, incorporation of 2-butyne between DoxSF and Tam would have been a better design for a construct targeted to ER.

Unlike ER, AR exists primarily as a cytosolic receptor in complex with several heat-shock proteins. Ligand binding leads to dissociation of the heat-shock proteins, homodimerization, and translocation into the nucleus where it promotes transcription of growth factors. Therefore, the binding of the construct to the cytosolic AR might logically lead to transport of the construct to the nucleus. Targeting and trafficking of DoxSF-Butyne-nilutamide in live prostate cancer cells was studied by fluorescence microscopy using AR negative PC-3 cells transiently transfected with the green fluorescent protein (GFP)-AR chimera [79]. The partial construct, SF-butyne-nilutamide, that has a binding IC$_{50}$ of 49 nM, caused the migration of the GFP-AR to the nucleus as monitored by movement of the GFP fluorescence in real time. The complete construct, DoxSF-butyne-nilutamide, bound to the GFP-AR as indicated from competition experiments but failed to cause the migration of the GFP-AR to the nucleus.

Drug localization and growth inhibition were compared in AR-positive PC-3/AR cells and AR-negative PC-3/neo cells. PC-3/AR and PC-3/neo cells are PC-3 cells permanently transfected with a vector bearing the gene for the AR gene and an empty vector, respectively. Anthraquinone fluorescence showed initial cytosolic localization of DoxSF-butyne-nilutamide, independent of AR expression. Possibly, an equivalent to the anti-estrogen binding site (AEBS) exists in prostate cancer cells to which the construct also binds. DoxSF-butyne-nilutamide also showed very similar growth inhibition of PC-3/neo cells and PC-3/AR cells and similar to DoxSF. Hence, in a static cell experiment, DoxSF-butyne-nilutamide showed no advantage over untargeted DoxSF. Although selectivity by DoxSF-Teg-Tam was observed over DoxSF with various breast cancer cells (Table 2), lack of selectivity of DoxSF-butyne-nilutamide for growth inhibition of PC-3 cells is possible because both DoxSF-butyne-nilutmide and DoxSF release Dox-formaldehyde conjugate at approximately the same rate. Possibly, DoxSF-butyne-nilutamide would show superiority under the dynamic conditions of an in vivo experiment.

Targeting $\alpha_V\beta_3$ integrin for metastatic cancer. The integrin $\alpha_V\beta_3$ is overexpressed on the surface of many tumor cells and endothelial cells responsible for angiogenesis, and its expression correlates with tumor progression in glioma, melanoma, breast cancer, and ovarian cancer. Several RGD peptides and peptide mimetics exhibit excellent binding to and selectivity for $\alpha_V\beta_3$. Our interest in using RGD peptides for targeting DoxSF to tumors and associated angiogenesis was initially stimulated by the work of Ruoslahti and coworkers [80]. Using in vivo phage display, they discovered a peptide, CDCRGDCFC (RGD-4C), that homed to tumor xenografts in mice. Furthermore, doxorubicin RGD-4C conjugates of undefined structure were effective in reducing tumor burden with less side effects. Structural studies of RGD-4C have shown that the RGD motif is rigidly displayed through a bicyclic ring system created by air oxidation of the four cysteins to two cystines [81]. Scheeren and coworkers synthesized a structurally well-defined RGD-4C doxorubicin conjugate with a plasmin cleavage site. This design incorporated two tumor targeting mechanisms, $\alpha_V\beta_3$ and plasmin, plus a mechanism for separating doxorubicin from the targeting groups at the site of the tumor. Plasmin is a serine protease overexpressed by metastatic tumor cells and vascular endothelial cells associated with their angiogenesis [82]. Our initial efforts were focused on DoxSF conjugates of RGD-4C but later turned to DoxSF conjugates of a simpler cyclic peptide, (N-Me-VRGDf), know as Cilengitide [83]. Cilengitide has proceeded as far as phase II clinical trials as a potent antagonist of $\alpha_V\beta_3$ [84]. Functionalization of Cilengitide for attachment of the tether was at the para position of the phenyl group of the d-phenylalanine ring. Again, the attachment point was directed by a co-crystal structure that showed the exposure of this position [85]. The structure of the construct, cyclic-(N-Me-VRGDf-NH)-DoxSF, is shown in Fig. 9. The complete construct maintained the high affinity for $\alpha_v\beta_3$ (binding IC$_{50}$, 5 nM) in a vitronectin cell adhesion assay relative to the peptide bearing only the tether (0.5 nM) [83]. However, the IC$_{50}$ for growth inhibition of MDA-MB-435 cells was 90 nM, approximately 2-fold larger than that of DoxSF. Flow cytometry and growth inhibition experiments suggested that the construct does not penetrate the plasma membrane but that the released doxorubicin-formaldehyde conjugate does. This may explain the higher IC$_{50}$ for growth inhibition relative to DoxSF because drug delivery is now limited by the abundance of receptors. Possibly, better growth inhibition would have been observed with cells that express more $\alpha_v\beta_3$ such as MDA-MB-231 cells.

The primary lesson learned from these initial studies of targeted doxorubicin-formaldehyde conjugates was that successful designs need to incorporate drugs with high activity because of the limited number of receptors. A secondary lesson is that tether design is important with linear tethers generally being superior. The primary lesson prompted us to design targeted drugs that used doxazolidine in place of DoxSF. Doxazolidine is at least an order of magnitude more active than DoxSF. The initial challenge was to discover a method to inactivate doxazolidine and stabilize its oxazolidine ring until the construct reached its target.

5
Doxazolidine (Doxaz), Targeted Doxazolidine, and Some Other Targeting Strategies for the Alkylating Anthracyclines

Why is doxoform at least an order of magnitude more active than DoxSF? To review what was introduced in Sect. 3, the reaction of Dox with formalin gives doxoform. An intermediate in the synthesis of doxoform is doxazolidine (Doxaz), and relatively pure Doxaz can be prepared by the reaction of Dox with paraformaldehyde in chloroform, stopping the reaction at the halfway point as determined by NMR spectroscopy [50]. Doxoform rapidly hydrolyzes in an aqueous medium at pH 7.4 and 37 °C to Doxaz (half-life, 1 min) and Doxaz rapidly hydrolyzes to Dox (half-life, 3 min). An intermediate in the latter hydrolysis is the acyclic doxorubicin-formaldehyde conjugate, the same species released by DoxSF. DoxSF releases the acyclic doxorubicin-formaldehyde conjugate with a half-life of 60 min [73]. Further, epidoxoform, like DoxSF, is an order of magnitude less active than doxoform; it hydrolyzes to an acyclic epidoxorubicin-formaldehyde conjugate with a half-life of 2 h [58]. Doxoform and doxazolidine inhibit the growth of cancer cells equally well. If we assume that the *virtual cross-linking* of DNA is the source of activity of doxorubicin-formaldehyde conjugates, then what is the mechanism of this cross-linking? Is the reactive species doxazolidine or the acyclic doxorubicin-formaldehyde conjugate? Does the doxorubicin-formaldehyde conjugate first intercalate in DNA and then form a covalent bond to the 2-amino group of a G-base? Or, does covalent bonding occur best in concert with the ring opening of the oxazolidine ring? The results of many experiments now point to covalent bond formation in concert with the ring opening as the most favorable way of creating the cross-link, as shown schematically in Fig. 11. Although epidoxoform cannot form the oxazolidine, DoxSF upon hydrolysis can form doxazolidine, but in competition with loss of formaldehyde, which may explain its lower activity. Molecular models suggest that a prelude to cross-linking by doxazolidine must be the tautomerization of the G-base to create a nucleophilic site at the 2-amino group. This is a simple but unexplored in-plane tautomerization that is likely important in cross-linking DNA by other drugs at G-bases in the minor groove such as by mitomycin C and FR-900482 [73]. This analysis suggests that the key for drug development is targeted delivery of doxazolidine in a stable, inactive form with release by an enzyme overexpressed at the tumor.

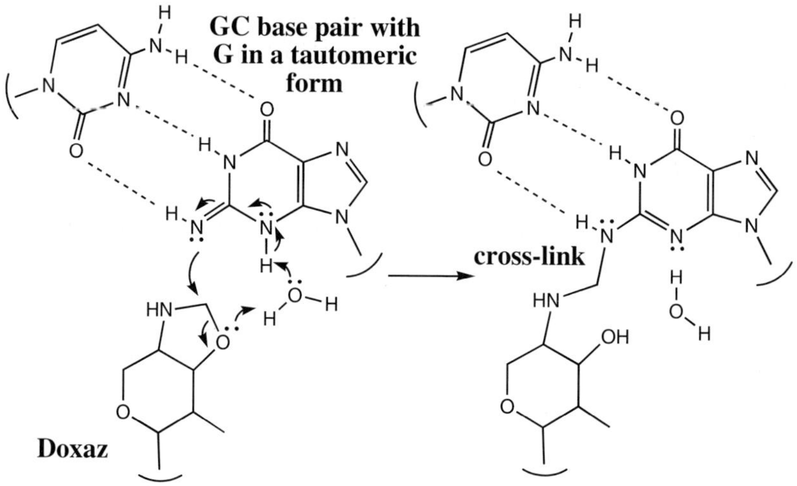

Fig. 11 A proposed reaction mechanism for direct covalent bonding of doxazolidine to a G-base of DNA. The reaction mechanism requires a prior tautomerization of the G-base to create an in-plane lone pair on the nitrogen at the 2-position of the G-base. As the covalent bond is forming, the anthraquinone part of the compound intercalates between stacked base pairs of the DNA. The model maintains the integrity of the electrons in the σ- and π-regions of space and leads to no separation of formal charge

Oxazolidine rings are stabilized with respect to the spontaneous loss of formaldehyde by acylation of the nitrogen. Acylation to create a carbamate offers the possibility for releasing doxazolidine through spontaneous decarboxylation of a carbamic acid from hydrolysis at the ester side of the carbamate. The idea of an enzyme-activated carbamate prodrug is stimulated by the relatively new clinical anti-cancer drugs Capecitabine and Irinotecan. Capecitabine is a prodrug that is activated to 5-fluorouracil by three enzymatic steps, the first of which is hydrolysis of a carbamate primarily by carboxylesterase 1 (CES 1, hCE1) [86–88]. Irinotecan is a prodrug that is activated to a water soluble camptothecan derivative by hydrolysis of its carbamate, primarily by carboxylesterase 2 (CES2, hiCE) [89, 90]. These drugs and their activation are shown in Fig. 12.

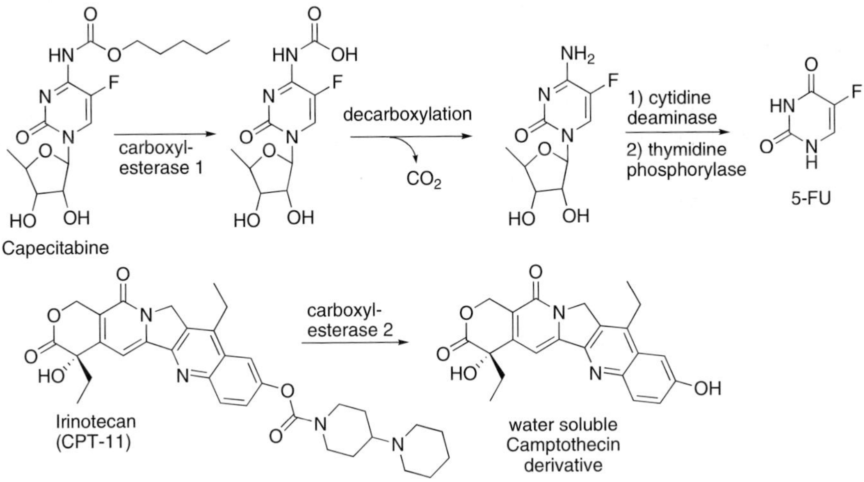

Fig. 12 Structures of the clinical prodrugs Capecitabine and Irinotecan, activated by carboxylesterase enzymes

After evaluating several designs, our lead design of a carboxylesterase-activated doxazolidine derivative incorporated a pentyl carbamate analogous to that of Capecitabine, separated from doxazolidine carbamate by a self-eliminating spacer [52]. The construct, pentyl-PABC-Doxaz, is shown in Fig. 13 together with its proposed activation to doxazolidine by a carboxylesterase enzyme. The design is for primary liver cancer because liver cancer cells overexpress CES 1. Colon cancer is also a possibility because CES 2 is overexpressed in colon cancer cells. IC_{50} values for growth inhibition of SK-HEP-1 and Hep G2 primary liver cancer cells are 1000 nM and 50 nM, respectively, with 24 h drug treatment. Corresponding numbers for treatment with Dox are 50 nM and 30 nM. Therefore, pentyl-PABC-Doxaz is of similar activity to Dox against Hep G2 cells but an order of magnitude less active than Dox against SK-HEP-1 cells. This is consistent with mRNA expression of CES1 and CES2 in the two cell lines as measured by RT-PCR [52]. Hep G2 cells express both CES1 and CES2, but SK-HEP-1 cells express some CES2 but little if any CES1.

Although pentyl-PABC-Doxaz only has similar activity to Dox against carboxylesterase-expressing cells, its activity against rat cardiomyocytes is

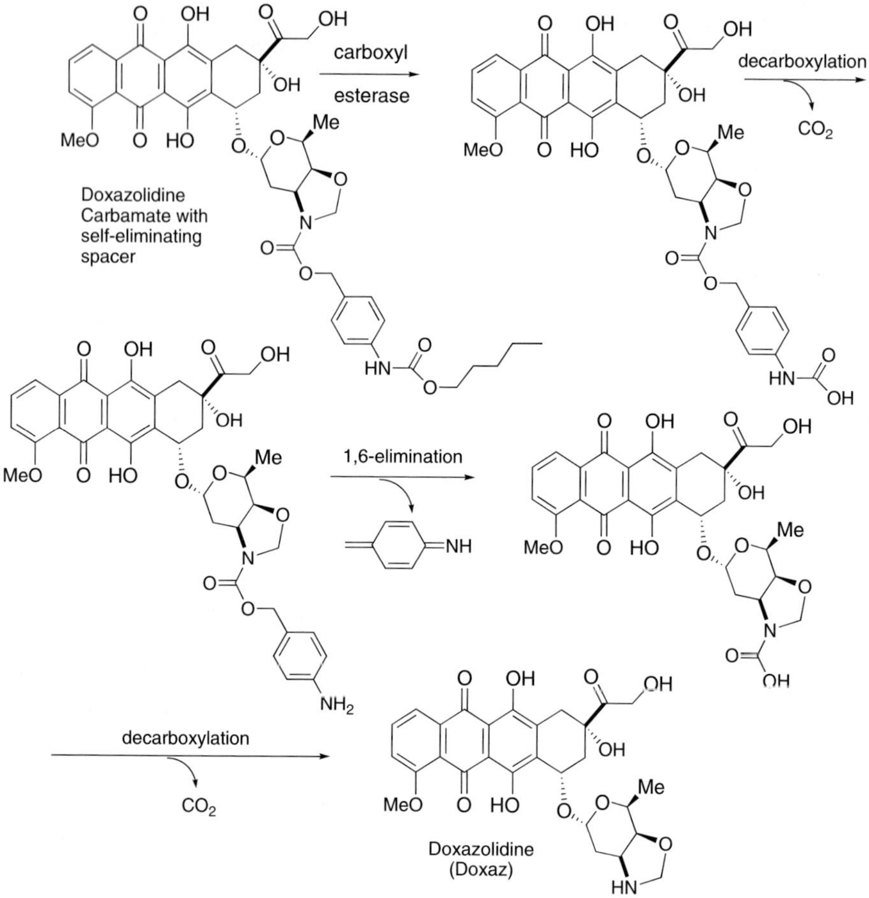

Fig. 13 Structure of doxazolidine prodrug, pentyl PABC-Doxaz, and mechanism of activation by carboxylesterase enzymes. The design utilizes a Katzenellenbogen self-eliminating spacer to separate the enzyme active site from the bulky cytotoxin

more than an order of magnitude less than that of Dox [52]. The IC_{50} values for pentyl-PABC-Doxaz and Dox are 630 and 20, nM, respectively. The growth inhibition of cardiomyocytes is a measure of cardiotoxicity. Together, the IC_{50} data predict an order of magnitude better safety factor for pentyl-PABC-Doxaz with carboxylesterase-expressing Hep G2 liver cancer cells versus cardiomyocytes.

Targeting strategies for other alkylating anthracyclines. Extensive research by Nagy, Schally and their coworkers has led to 2-pyrrolinodoxorubicin (AN-201) targeted to several important receptors overexpressed by cancer cells [91]. The targeting groups are derivatives of peptide hormones and are attached at the 14-position via a glutarate tether. Examples of the peptide hormones include somatostatin, luteinizing hormone-releasing hormone, and bombesine. The corresponding targeted 2-pyrrolinodoxorubicins have the acronyms AN-238, AN-152, and AN-215. As an example, the structure of AN-238 is shown in Fig. 14. Whereas 2-pyrrolinodoxorubicin is highly toxic to mice, the conjugates are much less toxic and effective in reducing the growth of a wide variety of tumor xenografts [91]. An important difference between the two alkylating anthracyclines, 2-pyrrolinodoxorubicin and doxazolidine, is their predicted lifetimes in vivo. Doxazolidine loses its alkylating functional group, formaldehyde, with a predicted half-life of several minutes, whereas the alkylating functional group of 2-pyrrolinodoxorubicin is robustly tethered to the 3′-amino group of doxorubicin. We view the labile nature of doxazolidine as a positive feature of targeted doxazolidine prodrugs because doxazolidine that escapes the tumor will hydrolyze within a few minutes to less toxic doxorubicin. This should help minimize some side effects of prodrug therapy with highly cytotoxic drugs.

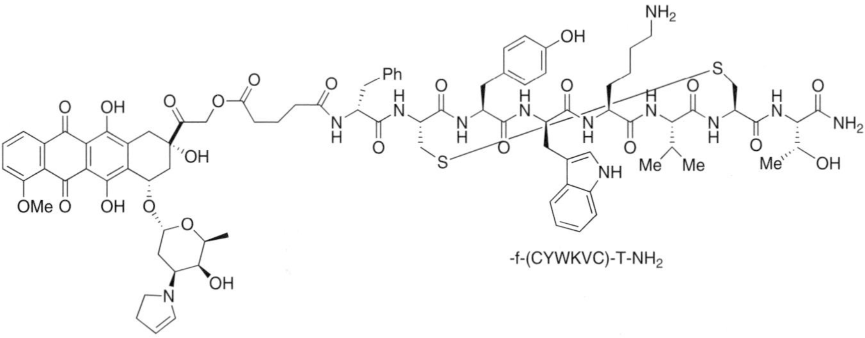

Fig. 14 Structure of 2-pyrrolinodoxorubicin tethered to somatostatin analog

6
Co-Administration of Doxorubicin and Formaldehyde Prodrugs

Yet another approach to facilitating DNA *virtual cross-links* in cancer cells is the combination of an anthracycline and a formaldehyde prodrug as promoted by Nudelman, Rephaeli, Phillips, and coworkers [92]. The most promising formaldehyde prodrug design is a bis-ester of formaldehyde hydrate as shown in Fig. 15 with one of the acid components being butyric acid, as illustrated by AN-9, AN-7, and AN-193. These compounds were actually designed to be prodrugs of butyric acid which is a histone deacetylase inhibitor. In this capacity, AN-9 has advanced to clinical trials [93, 94]. Early on, AN-9 exhibited synergy with daunorubicin in cell culture experiments and in cancer-bearing mice [95]. Subsequently, AN-9, together with doxorubicin, was shown to enhance DNA adduct formation in cancer cells with a synergistic effect on cancer cell growth [96]. Hexamethylene tetramine (Fig. 15),

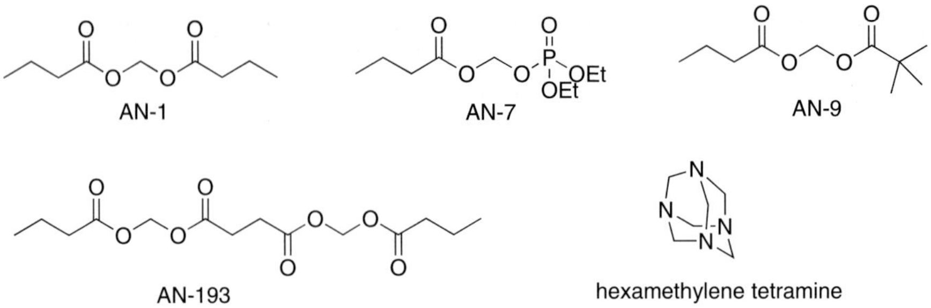

Fig. 15 Structures of formaldehyde prodrugs

which contains six formaldehyde equivalents, has also been explored as a formaldehyde-releasing prodrug, and co-administration with doxorubicin showed synergy with enhanced DNA adduct formation [97].

7
Summary and Conclusions

Early in the development of the anthracycline anti-tumor drugs, scientists recognized that DNA was an important drug target, and in vitro drug-DNA intercalation was established by crystallography. Drug intercalation is likely involved in the formation of DNA double strand breaks in vivo through its effect on the topoisomerase II decatenation of DNA, triggering cell death. Quinone redox activity was also observed early on and linked to the induction of oxidative stress and drug metabolism. The induction of oxidative stress has been implicated in drug cardiotoxicity. An intermediate in drug metabolism, the quinone methide from reductive glycosidic cleavage, was proposed as a possible electrophilic alkylating agent for DNA amongst other macromolecular targets, but it proved to be more nucleophilic than electrophilic. More recent studies now implicate formaldehyde as a mediator of anthracycline-DNA alklyation in vitro and in vivo. Possibly, the anthracycline-induction of oxidative stress leads to elevation of formaldehyde levels. Formaldehyde-mediated DNA-alkylation and intercalation at 5'-GC-3' sites *virtually cross-links* the DNA, significantly raising the stability of the duplex structure. By some mechanism, apparently independent of topoisomerase II, DNA-drug *virtual cross-links* trigger cell death.

A number of questions remain unanswered. What is the carbon source of the formaldehyde? In doxorubicin therapy, what fraction of tumor response stems from intercalation/topo II lesions and what fraction from *virtual-crosslinking*? How does *virtual crosslinking* trigger cell death and is it repairable?

The anthracyclines react with formaldehyde at their 3'-amino group to give conjugates that are hydrolytically unstable but show higher tumor cell growth inhibition than their clinical counterparts. A drug design goal has been a more stable compound that would release a reactive doxorubicin-formaldehyde conjugate in or near the tumor cell or immature endothelial cells associated with tumor angiogenesis. An early conjugate design employed the formaldehyde-N-Mannich base functional group in this capacity. The compound doxsaliform (DoxSF), has a half-life of 1 h with respect to the release of an acyclic doxorubicin-formaldehyde conjugate. Hence, the N-Mannich base serves as a time-release device. DoxSF was explored in the design of a targeted prodrug for various receptors overexpressed by tumor cells and/or their associated angiogenesis. The experiments indicated that DoxSF-like doxorubicin was not active enough for targeted drug design. The most active conjugates are doxoform and doxazolidine with doxoform being a dimeric prodrug of doxazolidine. In spite of their short half-lives with respect to hydrolysis to doxorubicin, they inhibit the growth of a broad spectrum of tumor cells by 1 to 4 orders of magnitude better than doxorubicin. Furthermore, they may help address the chronic side effect of doxorubicin cardiotoxicity because they are more toxic to cancer cells than to cardiomyocytes. In contrast, doxorubicin is more toxic to cardiomyocytes than cancer cells.

The systemic administration of doxazolidine is, however, problematic because doxazolidine is also toxic to some normal cells. For example, it inhibits the growth of Vero cells, green monkey kidney cells, by two orders of magnitude better than doxorubicin [52]. Furthermore, it will likely hydrolyze to doxorubicin with a half-life of a few minutes [50]. We propose that therapy with enzyme-activated doxazolidine prodrugs by enzymes overexpressed at the site of tumors will be advantageous because doxazolidine released at the tumor will be fast-acting and highly toxic to both drug-sensitive and drug-resistant tumor cells. It should also be active against tumor angiogenesis. Drug molecules that escape the site of activation will be the less toxic doxorubicin. The challenge will be the discovery of overexpressed enzymes and corresponding prodrugs. Current efforts are now focused on carboxylesterase-activated doxazolidine carbamate for primary liver cancer [52]. A complementary treatment strategy of possibly broader scope is the co-administration of doxorubicin with a formaldehyde-releasing prodrug such as AN-9 [92]. Yet a further challenge will be the discovery of drug/prodrug combinations that are synergistic.

Acknowledgements We thank the US NIH (R01 CA-92107) and US DOD (DAMD17-01-1-0046) for financial support. B.L.B. also thanks the US NIH for support under training grant T32 GM008759 to the University of Colorado.

References

1. Cutts SM, Nudelman A, Rephaeli A, Phillips DR (2005) IUBMB Life 5:73
2. Zunio F, Gambretta R, Di Marco A, Zaccara A (1972) Biochim Biophys Acta 277:489

3. Arcamone F (1981) Doxorubicin Anticancer Antibiotics. Academic Press, New York
4. Wang AH-J, Ughetto G, Quigley GJ, Rich A (1987) Biochemistry 26:1152
5. Frederick CA, Williams LD, Ughetto G, Vander Marel GA, Van Boom J, Rich A, Wang AH-J (1990) Biochemistry 29:2538
6. Chaires JB (1990) Biophys Chem 35:191
7. Chaires JB, Satyanarayana S, Suh D, Fokt I, Przewloka T, Priebe W (1996) Biochemistry 35:2047
8. Liu LF (1989) Ann Rev Biochem 58:351
9. Watt PM, Hickson ID (1994) Biochem J 303:681
10. Pommier Y (1995) In: Priebe W (ed) Anthracycline Antibiotics: New Analogues, Methods of Delivery, and Mechanisms of Action, vol ACS Symposium Series 574. American Chemical Society, Washington DC, p 183
11. Sinha BK, Trush MA, Kennedy KA, Mimnaugh EG (1984) Cancer Res 44:2892
12. Cummings J, Bartoszek A, Smyth JF (1991) Anal Biochem 194:146
13. Skladanowski A, Konopa J (1994) Biochem Pharmacol 47:2279
    Skladanowski A, Konopa J (1994) Biochem Pharmacol 47:2269
14. Swift LP, Rephaeli A, Nudelman A, Phillips DR, Cutts SM (2006) Cancer Res 66:4863
15. Nakazawa H, Riggs CE Jr, Egorin MJ, Redwood SM, Bachur NR (1984) J Chromatogr 307:323
16. Kleyer D, Koch TH (1984) J Am Chem Soc 106:2380
17. Gaudiano G, Koch TH (1991) Chem Res Toxicol 4:2
18. Moore HW, Czerniak R (1981) Med Res Rev 1:249
19. Egholm M, Koch TH (1989) J Am Chem Soc 111:8291
20. Bachur NR, Gordon SL, Gee MV (1977) Mol Pharmacol 13:901
21. Pan S, Pedersen L, Bachur NR (1981) Mol Pharmacol 19:184
22. Lown WJ, Chen JA, Plambeck JA, Acton EM (1982) Biochem Pharmacol 31:575
23. Land EJ, Mukherjee T, Swallow AJ, Bruce JM (1983) Arch Biochem Biophys 225:116
24. Doroshow JH (1983) Cancer Res 43:4543
25. Doroshow JH, Locker GY, Myers CE (1980) J Clin Invest 65:128
26. Goormaghtigh E, Chatelain P, Caspers J, Ruysschaert JM (1980) Biochem Pharmacol 29:3003
27. Minotti G, Menna P, Salvatorelli E, Cairo G, Gianni L (2004) Pharmacol Rev 56:185
28. Wang S, Konorev EA, Kotamraju S, Joseph J, Kalivendi S, Kalyanaraman B (2004) J Biol Chem 279:25535
29. Cullinane C, van Rosmalen A, Phillips DR (1994) Biochemistry 33:4632
30. Gewirtz DA (1999) Biochem Pharmacol 57:727
31. Taatjes DJ, Gaudiano G, Resing K, Koch TH (1996) J Med Chem 39:4135
32. Taatjes DJ, Gaudiano G, Resing K, Koch TH (1997) J Med Chem 40:1276
33. Wang AH-J, Gao YG, Liaw YC, Li YK (1991) Biochemistry 30:3812
34. Zhang H, Gao YG, van der Marel GA, van Boom JH, Wang AH (1993) J Biol Chem 268:10095
35. Zeman SM, Phillips DR, Crothers DM (1998) Proc Natl Acad Sci USA 95:11561
36. Podell ER, Harrington DJ, Taatjes DJ, Koch TH (1999) Acta Cryst D55:1516
37. Leng F, Savkur R, Fokt I, Przewloka T, Priebe W, Chaires JB (1996) J Am Chem Soc 118:4731
38. Taatjes DJ, Gaudiano G, Koch TH (1997) Chem Res Toxicol 10:953
39. Shiraishi H, Kataoka M, Morta Y, Umemoto J (1993) Free Rad Res Commun 19:315
40. Tamura H, Kitta K, Shibamoto T (1991) J Agric Food Chem 39:439
41. Maleski KA (2002) Exploration of possible sources of formaldehyde from anthracycline antitumor drug redox chemistry, MS Thesis, University of Colorado

42. Fiallo MML, Drechsel H, Garnier-Suillerot A, Matzanke BF, Kozlowski H (1999) J Med Chem 42:2844
43. Thorndike J, Beck WS (1977) Cancer Res 37:1125
44. Kato S, Burke PJ, Fenick DJ, Taatjes DJ, Bierbaum VM, Koch TH (2000) Chem Res Toxicol 13:509
45. Kato S, Burke PJ, Koch TH, Bierbaum VM (2001) Anal Chem 73:2992
46. Bagchi D, Bagchi M, Hassoun EA, Kelly J, Stohs SJ (1995) Toxicology 95:1
47. Spanel P, Smith D, Holland TA, Singary WA, Elder JB (1999) Rapid Commun Mass Spectrom 13:1354
48. Ebeler SE, Hinrichs SH, Clifford A, Shibamoto T (1992) Anal Biochem 205:183
49. Fenick DJ, Taatjes DJ, Koch TH (1997) J Med Chem 40:2452
50. Post GC, Barthel BL, Burkhart DJ, Hagadorn JR, Koch TH (2005) J Med Chem 48:7648
51. Post GC (2006) Design, synthesis, and biological evaluation of anthracycline-formaldehyde conjugates, PhD Thesis, University of Colorado
52. Burkhart DJ, Barthel BL, Post GC, Kalet BT, Nafie JW, Shoemaker RK, Koch TH (2006) J Med Chem 49:7002
53. Inouye S (1968) Chem Pharm Bull 16:1134
54. Weenen H, Maanen JM, Planque MM, McVie JG, Pinedo HM (1984) Eur J Cancer Clin Oncol 20:919
55. van der Vijgh WJ, Maessen PA, Pinedo HM (1990) Chemother Pharmacol 26:9
56. Goldin A, Venditti JM, Geran R (1985) Invest New Drugs 3:3
57. Bellino R, Cortese P, Danese S, De Sanctis C, Durando A, Genta F, Grio R, Giardina G, Katsaros D, Massobrio M, Richiardi G, Vicelli R (2000) Anticancer Res 20:4825
58. Taatjes DJ, Fenick DJ, Koch TH (1998) J Med Chem 41:1306
59. Taatjes DJ, Koch TH (2001) Curr Med Chem 8:15
60. Dernell WS, Powers BE, Taatjes DJ, Cogan P, Gaudiano G, Koch TH (2002) Cancer Invest 20:712
61. Acton EM, Tong GL, Mosher CW, Wolgemuth RL (1984) J Med Chem 27:638
62. Uchida T, Imoto M, Takahashi Y, Odagawa A, Sawa T, Tatsuta K, Naganawa H, Hamada M, Takeuchi T, Umezawa H (1988) J Antibiot 41:404
63. Moufarij MA, Cutts SM, Neumann GM, Kimura K, Phillips DR (2001) Chem Biol Interact 138:137
64. Nagy A, Armatis P, Schally AV (1996) Proc Natl Acad Sci USA 93:2464
65. Cherif A, Farquhar D (1992) J Med Chem 35:3208
66. Priebe W, Przewloka T, Fokt I, Ling Y-H, Perez-Soler R (2004) US Patent 6 680 300
67. Quintieri L, Geroni C, Fantin M, Battaglia R, Rosato A, Speed W, Zanovello P, Floreani M (2005) Clin Cancer Res 11:1608
68. Cullinane C, Phillips DR (1994) Biochemistry 33:6207
69. Perrin LC, Cullinane C, Kimura K-I, Phillips DR (1999) Nucleic Acids Res 27:1781
70. Fowler CR (2000) A survey of the N-Mannich base derivatives of the anti-tumor anthracycline drugs: synthesis, stability and cytotoxicity of novel anthracycline compounds, MS Thesis, University of Colorado
71. Johansen M, Bundgaard H (1980) Int J Pharm 7:119
72. Loudon GM, Almond MR, Jacob JN (1981) J Am Chem Soc 103:4508
73. Cogan PS, Fowler CR, Post GC, Koch TH (2004) Lett Drug Design Dis 1:247
74. Kedjouar B, de Medina P, Oulad-Abdelghani M, Payre B, Silvente-Poirot S, Favre G, Faye J-C, Poirot M (2004) J Biol Chem 279:34048
75. Shiau AK, Barstad D, Loria PM, Cheng L, Kushner PJ, Agard DA, Greene GL (1998) Cell 95:927
76. Burke PJ, Koch TH (2004) J Med Chem 47:1193

77. Burke PJ, Kalet BT, Koch TH (2004) J Med Chem 47:6509
78. Cogan PS, Koch TH (2003) J Med Chem 46:5258
79. Cogan PS, Koch TH (2004) J Med Chem 47:5690
80. Arap W, Pasqualini R, Ruoslahti E (1998) Science 279:377
81. Assa-Munt N, Jia X, Laakkonen P, Ruoslahti E (2001) Biochemistry 40:2373
82. de Groot FMH, Broxterman HJ, Adams HPHM, van Vliet A, Tesser GI, Elder-kamp YW, Schraa AJ, Kok JR, Molema G, Pinedo HM, Scheeren HW (2002) Mol Cancer Therap 1:901
83. Burkhart DJ, Kalet BT, Koch TH (2004) Mol Cancer Therap 3:1593
84. Dechantsreiter MA, Planker E, Matha B, Lohof E, HoIzemann G, Jonczyk A, Good-man SL, Kessler H (1999) J Med Chem 42:3033
85. Xiong J-P, Stehle T, Zhang R, Joachimiak A, Frech M, Goodman SL, Arnaout MA (2002) Science 296:151
86. Shimma N, Umeda I, Arasaki M, Murasaki C, Masubuchi K, Kohchi Y, Miwa M, Ura M, Sawada N, Tahara H, Kuruma I, Horii I, Ishitsuka H (2000) Bioorg Med Chem 8:1697
87. Walko CM, Lindley C (2005) Clin Therap 27:23
88. Quinney SK, Sanghani SP, Davis WI, Hurley TD, Sun Z, Murry DJ, Bosron WF (2005) J Pharmacol Exp Therap 313:1011
89. Bencharit S, Morton CL, Howard-Williams EL, Danks MK, Potter PM, Redinbo MR (2002) Nat Struct Biol 9:337
90. Sanghani SP, Quinney SK, Fredenburg TB, Davis WI, Murry DJ, Bosron WF (2004) Drug Metab Dispos 32:505
91. Nagy A, Schally AV (2005) Curr Pharm Design 11:1167
92. Cutts SM, Swift LP, Rephaeli A, Nudelman A, Phillips DR (2005) Curr Med Chem: Anti-Cancer Agents 5:431
93. Nudelman A, Ruse M, Aviram A, Rabizadeh E, Shaklai M, Zimra Y, Rephaeli A (1992) J Med Chem 35:687
94. Reid T, Valone F, Lipera W, Irwin D, Paroly W, Natle R, Sreedharan S, Keer H, Lum B, Scappaticci F, Bhatnagar A (2004) Lung Cancer 45:381
95. Kasukabe T, Rephaeli A, Honma Y (1997) Br J Cancer 75:850
96. Cutts SM, Rephaeli A, Nudelman A, Hmelnitsky I, Phillips DR (2001) Cancer Res 61:8194
97. Swift LP, Cutts SM, Rephaeli A, Nudelman A, Phillips DR (2003) Mol Cancer Ther 2:189

Top Curr Chem (2008) 283: 171–189
DOI 10.1007/128_2007_1
© Springer-Verlag Berlin Heidelberg
Published online: 6 November 2007

Sabarubicin

Federico-Maria Arcamone

Menarini Ricerche, Via Livornese 893, 56010 Pisa, Italy
farcamone@virgilio.it

Abstract Disaccharide derivatives in the daunorubicin and in the 4-demethoxy (idaru-bicin) series in which the first sugar moiety linked to the aglycone is a non-aminated sugar, namely 2-deoxy-L-rhamnose or 2-deoxy-L-fucose and the second moiety is daunosamine, have been obtained upon synthesis of the appropriate activated sugar intermediate and glycosylation of the corresponding aglycones. The compounds containing 2-deoxy-L-fucose exhibit superior pharmacological properties with respect to the stereoisomers containing 2-deoxy-L-rhamnose. The doxorubicin analog 7-O-(α-L-daunosaminyl-α(1–4)-2-deoxy-L-fucosyl)-4-demethoxy-adriamycinone (sabarubicin) was prepared starting from 14-acetoxyidarubicinone. Solution properties and molecular interactions are compared with those of doxorubicin. Sabarubicin exhibits a superior antitumor efficacy, presumably related to the activation of p53-independent apoptosis. A number of sabarubicin analogues have also been synthesized.

Keywords Demethoxyadriamycinone · Disaccharide · DNA · Glycosylation · Sabarubicin

1
Introduction

Anthracycline glycosides represent a wide class of natural compounds obtained by submerged aerobic fermentation of different microorganisms belonging to the genus *Streptomyces*. Their aglycone moieties, the anthracyclinones, are characterized by a tetracyclic system bearing an anthraquinone chromophore [1]. The anthracyclinones belong to the large family of the polyketide natural products, and are biosynthesized by the action of multi-functional polyketide synthase enzymes through repeated condensations of simple acylthioesters. The growing carbon chain containing β keto groups that is formed undergoes a number of intra-molecular reactions and a series of reductive steps leading ultimately to aromatic derivatives, such as the an-thracyclinones [2]. The final products are characterized by varied molecular structures, further diversified by the different glycosylation patterns typical for the different families of final glycosides. In the case of the anthracyclines, the pattern of glycosylation is a consequence, both in terms of the nature of the sugar moieties as well as of the position of the substitution, of the genetic constitution of the producing microbial strains [3–6]. The antitumor activity of a biosynthetic anthracycline and its chemical relationship with the glyco-sidic pigments previously studied and described by different authors was first reported in 1959 [7].

Doxorubicin (1, Adriamycin), the best known biosynthetic antitumor an-thracycline, is the 14-hydroxylated derivative of daunorubicin (2), the main product of the *Streptomyces peucetius* fermentation [8]. Since its registration in the early 1970s, 1 has been one of the most widely used drugs in can-cer chemotherapy for more than 30 years. The successful use of 1 in the medical treatment of cancer was followed by a wide effort aimed at the syn thesis of better analogues by chemical modification of the parent drugs or by total synthesis. The compounds epirubicin (3) and idarubicin (4), import-ant new members of this chemical group, are presently used in the medical practice. Compound 3 has been registered in most countries as Farmorubicin or Pharmorubicin. In the US, 3 is currently marketed as Ellence. Idaru-bicin is marketed with the trademark Zavedos and is used mainly in acute leukaemias. These compounds are currently described as second generation antitumor anthracyclines [9, 10]. Together with doxorubicin, epirubicin and idarubicin are presently also available as "generics". Antitumor activity has been associated with a number of other biosynthetic anthracyclines differing from the daunorubicin-doxorubicin group in the chemical structure of both the aglycone moiety and the sugar residue(s), the latter also present as an

1: R=OH
2: R=H

3

4

oligosaccharide fragment. Aclacinomycin (Aclacin), a biosynthetic trisaccharide derivative, has been introduced in clinical use [11].

A dose limiting factor in the repeated treatment with doxorubicin and related compounds is the development of cardiotoxicity. Also, a number of tumors of clinical importance, including colon, lung, pancreatic and renal cancers and malignant melanoma, do not respond to currently available anthracycline drugs. Other diseases, such as gastric and small cell lung cancers, and advanced ovary and breast tumors, are only partially responsive and the benefit of drug treatment is marginal. As for the clinical mechanism of resistance, only the classic multidrug resistance (MDR) phenotype, which is due to the presence of P-glycoprotein (PGP) in the plasma membrane (a "pump" that can extrude a wide range of anticancer drugs and other foreign compounds), has been shown to contribute to resistance in patients. Evidence for other possible mechanisms of resistance in clinical patients is not presently available [12].

Hundreds of analogues with considerable structural variations were synthesized and tested using murine models of transplantable leukaemias. This approach may not have allowed an appropriate identification of more selective new compounds with an improved therapeutic index over 1, or one with a substantially different spectrum of activity [13]. A more recent effort based on the search for a higher efficacy at the main molecular target, coupled with the use of a panel of human tumor cell lines and tumor type oriented pharmacology in laboratory animals, has resulted in new compounds closer to the desired goal [14]. In this chapter we shall describe the rationale that motivated the synthesis of the disaccharide analogues and the synthetic work leading to these third generation antitumor anthracyclines together with the molecular and biological properties and the available results concerning the presently available clinical data of the compound (laboratory code MEN 10755) selected for full development. "Sabarubicin" is the generic name proposed by WHO for this compound.

2
Chemistry of Sabarubicin

2.1
Design of the Disaccharide Doxorubicin Analogues

The understanding of the molecular mechanism of cell toxicity of antitumor anthracyclines is based on studies concerning (1) their binding to DNA involving an intercalation at pyrimidine-purine steps with a preference for CG accompanied by the binding of the sugar moiety within the minor groove [15–20], and (2) the association of the breaks of cell DNA induced by doxorubicin with the enzyme topoisomerase II, a good correlation being recorded between intensity of topoisomerase II mediated DNA breakage and cytotoxicity induced by different anthracyclines [21, 22]. The reaction of "topoisomerase poisoning" shows an outstanding selectivity because all sites of doxorubicin blockade display a specific requirement of an adenosine at a 3′ terminus which is never the case for the enzyme-only sites [23]. Moreover, the strength of DNA binding does not correlate with the stimulatory effect of anthracyclines on topoisomerase II-mediated DNA cleavage suggesting that the specific mode of DNA interaction, rather than the strength of binding, is important in determining the cytotoxic potency. The first step of the molecular mode of action of antitumor anthracyclines would be the formation of a drug-DNA-enzyme ternary complex, in which the enzyme is covalently linked to the broken DNA strand. The protein associated double-strand breaks that follow represent a DNA damage that triggers apoptotic cell death [24, 25]. In particular, the non-intercalating portion of the anthracycline molecule, namely the sugar moiety, is expected to critically influence the stability of the ternary complex, since in the latter the drug should be positioned at the interface of the active site of the enzyme and the DNA cleavage site.

As for the molecular mechanism of cardiotoxicity, the important side effect of the repeated dosages in responding patients, current evidence [26] is consistent with a fundamental role for free radicals, likely related to the presence of an oxidative stress caused by a higher than normal oxygen tension in the heart tissue [27]. Excess oxygen could be the result of the long lasting inhibition of nucleic acid synthesis [28] and consequent reduction of metabolic consumption of oxygen. This could be an effect related to the DNA interaction of the anthracyclines, as would the inhibition by doxorubicin of mitochondrial DNA transcription, leading to a diminished ATP production and consequently to cell damage in the high energy demanding tissue [29]. On the other hand, 1 selectively inhibits gene expression in cardiac muscle cells in vivo [30].

The substitution of a hydroxyl group with a hydrogen atom at positions 6 and/or 11 onto the aglycone moiety, although bearing noticeable consequences on the polarographic behaviour or on the chemical outcome of anaerobic quinone reduction, did not result in a marked change of bioactivity [31] as might have been the case if bioactivity of antitumor anthracyclines were related with the redox behaviour and/or with the ability to form metal complexes. A significant structural modification in terms of structure activity relationships was instead the substitution of the C-9 hydroxyl group on ring A with a hydrogen atom. In fact 9-deoxydaunorubicin, although showing the same affinity for DNA and similar electrochemical behaviour as the parent daunorubicin, was two orders of magnitude less cytotoxic on cultured HeLa cells. Therefore, it was deduced that the 9-hydroxyl group, which, according to the well known molecular model of the daunorubicin-DNA complex, protrudes outside the double helix, is involved in a specific interaction relevant to the bioactivity of the drug in terms of topoisomerase II blockade in the ternary complex [32].

In conclusion, as suggested by the DNA binding geometry and by the topoisomerase related mechanism of action of antitumor anthracyclines, a rationale design of new analogues would be based on the hypothesis that it should be beneficial to expand the molecular recognition properties through a functional enrichment of the portion of the anthracycline molecule involved in the stabilization of the "cleavable ternary complex". Structure activity relationships pointed to ring A as a "scaffold" determining the spatial orientation of the C-7 and C-9 substituents, in particular that of the sugar moiety, whose structure and stereochemistry is critical for the stabilization of the ternary complex [33].

Daunosamine, 3-amino-2,6 dideoxy-L-*lyxo*-hexose is the natural aminosugar present in all anthracycline glycosides produced by the strains belonging to the species *Streptomyces peucetius*, the original producer of the daunorubicin group of antibiotics, and related strains [3]. It has never been identified as a constituent of other anthracyclines nor of any natural glycoside. An early approach to new analogues consisted in the synthesis of new daunomycinone and adriamycinone aminoglycosides differing in the stereochemistry or in the structure from the biosynthetic daunosaminide, with the aim of modifying the pharmacokinetics and/or the metabolism of the natural glycosides. This approach afforded a number of compounds including the already-mentioned clinically useful 3, that exhibits the aminosugar acosamine, namely 3-amino-2,6-dideoxy-L-*arabino*-hexose.

However, the 3'-amino group is not a requisite for antitumor activity, as it is shown by the biological properties of the 3'-hydroxylated analogue of 1 [34]. Therefore, attention may be given to analogues containing two sugar residues, namely disaccharides in which daunosamine was not the first sugar moiety linked to the aglycone. A disaccharide, 4'-O-daunosaminyldaunorubicin, was present in the medium of the daunorubicin fermentation and the same, as well as the corresponding analogues in which the second sugar was either acosamine or 2-deoxy-L-rhamnose, were obtained by semisynthesis and found to be still bioactive, albeit somewhat less potent against murine transplantable leukaemias than daunorubicin [8]. The disaccharide derivative 4'-O-(2-deoxy-L-fucosyl)-daunorubicin has been synthesized upon reaction of daunomycinone with 1-O-acetyl-4-O-(2-deoxy-3,4-di-O-acetyl-L-fucosyl)-N-trifluoroacetyldaunosamine in the presence of p-toluenesulphonic acid, followed by basic deprotection of the product [35]. This disaccharide derivative has also been obtained from the same reagents using tin tetrachloride as the catalyst of the glycosidation [36]. Interestingly, a disaccharide analogue bearing a non-aminated sugar directly linked to the aglycone, namely 7-O-(3-O-α,L-daunosaminyl-2-deoxy-α,L-rhamnosyl)-daunomycinone, was synthesized and found biologically inactive [37].The lack of bioactivity should be related with the position of the second glycosidic linkage.

As for the substitution onto the aglycone moiety, the results of in vitro preclinical studies indicate that the higher lipophilicity of 4-demethoxy derivative **4** leads to a faster accumulation in the cell nuclei and consequently to enhanced cytotoxicity as compared to **2**. A major advantage over the 4-methoxylated parent drug is the ability to partially overcome multi-drug resistance. It has also been found that the major human metabolite of **4**, idarubicinol, is as bioactive as the parent compound [38].

2.2
Synthesis of 4′-(α,L-Daunosaminyl)-2′,6′-Dideoxyhexopyranosyl-Daunomycinone and -Idarubicinone with L-*Arabino* and L-*Lyxo* Configurations

Disaccharide derivatives of the daunorubicin and of the 4-demethoxy (idarubicin) series in which the first sugar moiety linked to the aglycone was a non-aminated sugar, namely 2-deoxy-L-rhamnose or 2-deoxy-L-fucose, and the second moiety was daunosamine, were synthesized upon glycosylation of the corresponding aglycones with the appropriate protected and activated disaccharides [39]. In the first series, the synthesis was started from 3,4-di-O-acetyl-L-rhamnal **5**, that was converted to p-methoxybenzyl 2-deoxy-α-L-rhamnoside **6** by reaction with p-methoxybenzyl alcohol and N-iodosuccinimide in acetonitrile at –50°, followed by dehalogenation with tributyltin hydride and then deacetylation with sodium methoxide in methanol. Selective p-nitrobenzoylation of an intermediate stannylene cyclic acetal of **6** afforded **7** that was converted, upon reaction with 1,4-di-p-nitrobenzoyl-3-N-trifluoroacetyl-L-daunosamine **8** in the presence of trimethylsilyltriflate, to **9a** in high yield and with the desired α configuration. The anomeric p-methoxybenzyl group was then removed with ceric ammonium nitrate and the free anomeric position of **9b** was esterified with p-nitrobenzoyl chloride to give the desired activated intermediate **9c**.

Similarly, 3,4-di-O-acetyl-L-fucal **10** was allowed to react with p-methoxybenzyl alcohol in the presence of a Lewis acid to give **11** that was in turn deacetylated with sodium methoxide in methanol and the resulting product uct selectively *p*-nitrobenzoylated to **12**. The difficulty of obtaining acceptable yields in the glycosylation of the axial hydroxyl group of **12** was overcome using the activated daunosamine derivative **13** and iodonium dicollidine perchlorate (IDCP). The disaccharide derivative **14a** was then converted to **14b** and finally to **14c** as described above for **9c**.

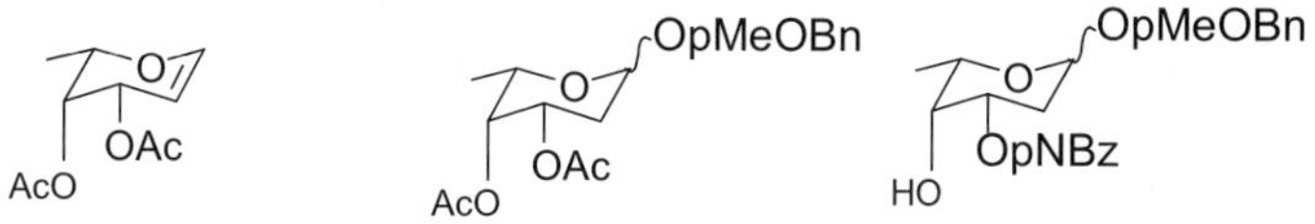

10 **11** **12**

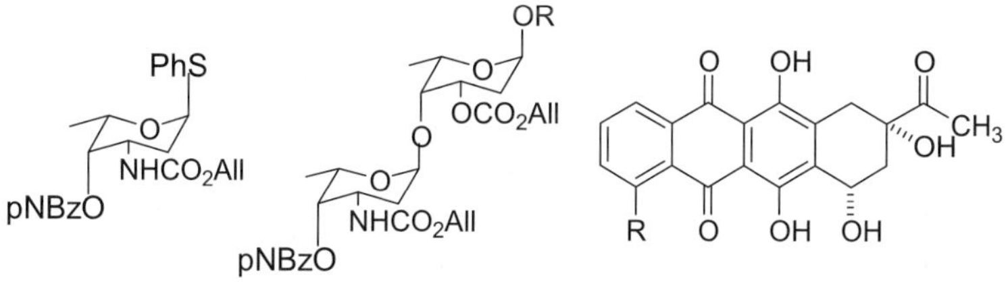

13

14a (R=pMeOBn)
14b (R=H)
14c (pNBz)

15a: R=OCH3
15b: R=H

The coupling reaction of both activated disaccharide units **9c** and **14c** with aglycones **15a** and **15b** was carried out using a standard procedure to afford, after deprotection with base, respectively the 4′-epimeric compounds **16a** and **17a** using **9c** as the glycosylating agent, and **16b** and **17b** using **14c**. Disaccharide glycosides containing 2-deoxy-L-fucose **17a,b** exhibited superior pharmacological properties with respect to the stereoisomers **16a,b** containing 2-deoxy-L-rhamnose [40].

2.3
Synthesis and Molecular Properties of Sabarubicin

On the basis of the results concerning the disaccharide analogues of daunorubicin and idarubicin, the doxorubicin analogue 7-*O*-(α-L-daunosaminyl-α(1–4)-2-deoxy-L-fucosyl)-4-demethoxyadriamycinone (sabarubicin, **18**) was eventually synthesized. Protected derivative **23** was prepared from 14-acetoxyidarubicinone **17** that was glycosylated with the activated disaccharide **21b**, in turn obtained from **21a** through the substitution of the protecting *p*-methoxybenzyl group with the *p*-nitrobenzoyl residue *via* reaction with the Ce(NH$_4$)$_2$(NO$_3$)$_6$ reagent followed by *p*-nitrobenzoylation of the free anomeric position. Compound **21a** was prepared upon condensation of the

16a: R=H
16b: R=OMe

17a: R=H
17b: R=OMe

N-allyloxy and *O*-allyloxy intermediates **19** and **20** in the presence of IDCP. Stepwise removal of the protecting groups of **23** followed by a reverse phase chromatographic purification and final lyophilization allowed **18** as the hydrochloride in good yield [41].

A PMR study, supported by molecular mechanical calculations, of the preferred conformation of sabarubicin in deuterated dimethylsulfoxide solution has been performed [42]. Experimental data showed the presence of an aggregation process attributed to the vertical stacking of the anthraquinone chromophore, in which the conformation of **18** may be conceived as a preferred physical state of the molecule, reasonably similar to that adopted in the drug-DNA intercalation complex [43]. The conformation of the aglycone saturated ring A was established as the $^9H_8$ half chair typical of the bioactive antitumor anthracyclines on the basis of the relevant coupling constant values for H-7, H-8 eq. and H-7, H-8 ax. of, respectively, 2.5 and 4.9 Hz, as well as of the long distance coupling of H-8 eq. and H-10 eq. equal to 2 Hz. The fucose and the daunosamine rings are in the expected $^1C_4$ chair conformation as deduced by the values of the coupling constants and by the NOE data. The establishment of the most stable conformation of the molecular system represented by **18** needed however the assignment of the geometry at the glycosidic bonds of the two sugar moieties, defined by the values of the angles $\phi = C(7) - O(7) - C(1') - H(1')$, $\psi = H(7) - C(7) - O(7) - C(1')$, $\phi' = C(4') - O(4') - C(1'') - H(1'')$, and $\psi' = H(4') - C(4') - O(4') - C(1'')$. A systematic search for the values of the said angles through molecular mechanics computations allowed a set of conformers that were analysed according to a modified version of the NAMFIS program [44] on the basis of the NOE and $^3J_{CH}$ data. The major conformer showed values of dihedral angles ϕ and ψ, respectively, 42° and −15°, corresponding to an orientation of the sugar residue directly linked to the aglycone moiety not different from that of the parent compound, doxorubicin. However, up to a total of four conformations were deduced from this study. In fact, the presence of a significant population of conformers in which ϕ and ψ were respectively 37° and 131° could not be excluded, whereas ϕ' and ψ' were attributed values of 49° and 6° (for a main conformer) and 89° and 24° (for a minor conformer), thus defining the shape of different species.

Sabarubicin is subjected to the same protonation equilibria as is doxorubicin [45]. According to the electronic and fluorescent spectra in buffered aqueous solutions and the results of potentiometric titrations, the behaviour at different pH values is similar, taking account of the absence, in sabarubicin, of the methoxy group at C-4. The self-aggregation process due to the stacking of the anthraquinone chromophores was characterized as a dimerizaton with association constant $\kappa \approx 2850\ \mathrm{M}^{-1}$, a value distinctly lower than that observed for the dimerization constant of doxorubicin ($\kappa \approx 23\,000\ \mathrm{M}^{-1}$).

2.4
Interactions with Biological Macromolecules

Spectrophotometric and fluorescence titrations of **18** with calf thymus DNA [45] and with the self-complementary d(CGATCG) hexamer [46] revealed spectral changes implying a similar binding mechanism and stability of the resulting complexes in line with what is already known for the parent drug doxorubicin. The presence of the second sugar residue seems to be irrelevant in the intercalation process. This involves the positioning of the planar chromophore between two adjacent base pairs and van der Waals interactions of the sugar moiety directly linked to the aglycone with structural elements within the minor groove of the double helix. However, the X-ray diffraction studies carried out on the orange-red crystals of the sabarubicin-d(CGATCG) complex support the anticipation that the second sugar residue of the disaccharide moiety might interact with other cellular components in close proximity with DNA. The asymmetric unit of the crystals contains one oligomeric duplex, two bound drug molecules and 35 water molecules. Similarly to the complexes of the other antitumor anthracyclines with the same oligodeoxynucleotide, intercalation occurs in the CpG steps at each end of the DNA duplex with the drug chromophoric system perpendicular to the long axis of the duplex, and a certain distortion of the B-type helix is generated. Varying from the already studied complexes, the two drug molecules have different conformations and the two binding sites show large differences indicating the possibility of different binding modes. In one site, the first sugar is in the chair conformation with the 3'-hydroxyl group still in van der Waals contact with the guanine residue and is rotated by about 45° around the O7-C1' bond with respect to the corresponding arrangement in the doxorubicin complex, whereas the second sugar moiety protrudes outside the double helix and the amino and hydroxyl groups interact with two water molecules. In the other site, both sugar rings lie in the minor groove, the 3'-hydroxyl of the fucose residue being in van der Waals contact with a cytosine, and the second sugar is in the boat conformation. An important interaction between the amino group of the disaccharide moiety bound to the first site and a guanine residue of a second DNA molecule, different from the one where the drug is intercalated, is observed. Actually, in the crystal lattice two layers of the stacked duplexes formed by the drug-DNA complexes are arranged mutually perpendicular to one another generating packing contacts, two of which involve the drug molecule. These are a hydrogen bond of the C-14 hydroxyl group towards a phosphate oxygen of the crossing complex and the other is the interaction mentioned above, thus providing the first example of an anthracycline-DNA complex where a cross-link with a second DNA helix has been observed (Fig. 1).

A pharmacokinetic study in human patients indicated relatively high plasma levels of sabarubicin and a much smaller volume of distribution in the body as compared with doxorubicin or epirubicin [47]. This observation raises interest as to the ability of the drug to bind human serum albumin (HSA), a major constituent of the serum proteins. In fact it has been found that the binding of sabarubicin to HSA is stronger than that shown by doxorubicin by two orders of magnitude [48]. Complexation has profound effects on the spectroscopic behaviour of the drug, recorded as modification of the electronic spectrum and as a marked quenching of the typical fluorescence of the anthracycline chromophore. Using visible and fluorescence measurements

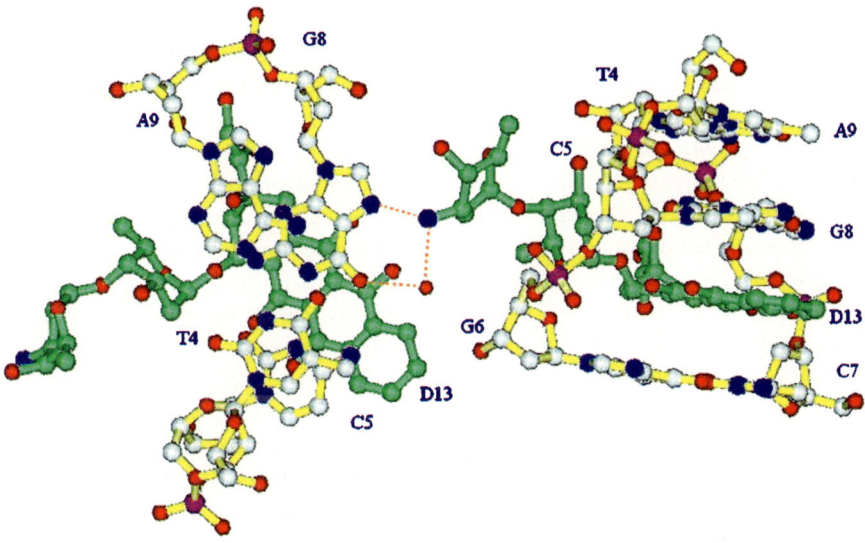

Fig. 1 Crystal structure of the sabarubicin–d(CGATCG) complex showing the interaction of the intercalated drug with a second DNA molecule. In DNA the nitrogen atoms are blue, the oxygen atoms are red, the phosphorous atoms are *purple* and the carbon atoms are *grey*. The drug molecule is *green*. Reproduced from [46] with permission from the publisher

the authors were able to determine the value of the corresponding binding constant as $1.1(\pm 0.3) \times 10^5$. The reversible character of the association was also demonstrated because a full recovery of the fluorescence was obtained upon dialysis of the complexes. The high affinity site was explored by ligand competition experiments. The conclusion was that **18** does not interact with the classical drug binding sites of HSA but shares a non-canonical binding site with ethacrynic acid. Taking account of the high concentration of HSA in plasma, it can be deduced that binding of **18** to HSA may provide an explanation not only of the unique pharmacokinetic behaviour, but also of the varied pharmacokinetics of the compound recorded in the clinical setting, a variation not found for other antitumor anthracyclines.

3
Pharmacology and Therapeutic Application

3.1
Preclinical Studies

It is well known that apoptosis, i.e. programmed cell death, is a common mode of cell death for tumor cells exposed to pharmacological doses of different cytotoxic agents. This has also been established for the antitumor anthracyclines, whose apoptotic response has been related to the generation of DNA breaks due to the formation of stable drug-DNA-topoisomerase II complexes. The enhanced potency against the topoisomerase II reaction exhibited by sabarubicin, as compared with doxorubicin, explains the similar extent of drug induced DNA breaks in human ovarian carcinoma cells in vitro when the cells are exposed to equal external concentrations of the two drugs, notwithstanding the lower intracellular concentration of the former compound. Despite a distinctly reduced rate of cell uptake, sabarubicin inhibited by 50% cultured A2780 human ovarian carcinoma cells at a molar concentration in the external medium equal to 0.39 µM, a value very close to that shown by doxorubicin of 0.40 µM. The high activity of sabarubicin has been tentatively related to the activation of p53-independent apoptosis. When tested in experimental tumor models, sabarubicin exhibited a superior antitumor efficacy when compared with doxorubicin. A very significant activity was found against human tumors such as A2780 ovarian tumor, MX-1 breast carcinoma, and POVD small cell lung cancer xenografted in athymic nude mice, the compound appearing markedly superior to doxorubicin in inhibiting tumor growth and in terms of an increased number of disease-free survivors among treated animals [41]. Clearly, the tumor models used were significantly different from naturally occurring human diseases for a number of reasons, including the difference in species between tumor and host, the absence of normal immunological reactions, and the specific site of implantation. The results were however indicative of a different, potentially favorable pharmacological behaviour of sabarubicin with respect to the main clinically useful anthracycline.

The in vitro and in vivo preclinical evaluation has been extended to a range of human tumors. In tumor cell cultures sabarubicin showed similar or lower cytotoxicity than doxorubicin. On the other hand, when tested against human tumor xenografts, sabarubicin was effective in 11/16 tumor cell lines (doxorubicin 5/16), and responsive cell lines included tumor cell lines originated from the lung or the prostate and of the gynecological type (breast and ovary). Differing from doxorubicin, sabarubicin was found to induce phosphorylation, and therefore inactivation, of the antiapoptotic factor bcl-2 that is over-expressed in the MX-1 tumor xenografts (as well as in other tumors), explaining the higher apoptotic properties of the disaccharide derivative [49, 50].

The pharmacokinetic properties of sabarubicin were studied in more detail and in comparison with doxorubicin using the ovarian carcinoma A2780 cell line in vitro and in vivo. The results of the in vitro experiments with the ^{14}C-labelled compound confirmed the reduced cell uptake together with a higher cytoplasmic/nuclear concentration ratio in comparison with doxorubicin. Also, the greater stimulation of DNA cleavage was associated with a longer persistence of the DNA breaks, very likely a consequence of a higher stability of the drug-DNA-topoisomerase II ternary complex [51]. In the comparative study in athymic nude xenografted mice drug concentrations were determined by HPLC with fluorometric detection. After a single 7 mg/kg intravenous injection of the drugs, sabarubicin showed a significantly lower accumulation in all tissues investigated, albeit less so in the tumor while, again, an enhanced antitumor efficacy was accompanied by a more marked activation of apoptosis [52].

In addition to a superior antitumor activity as compared with doxorubicin, sabarubicin proved to be less cardiotoxic in the rat. In fact, when repeated treatments with equimyelotoxic doses of the two agents where tested, the disaccharide exhibited lower cardiotoxic responses including less severe myocyte lesions. Moreover, different from that observed with doxorubicin, functional and histological effects were not progressive after the end of treatment [53]. In agreement with the results of this study are the observations that sabarubicin exhibited much less effect on Ca^{2+} channels of rat heart sarcoplasmic reticulum and on ryanodine receptor than doxorubicin. It was also seen that the presence of 30 µM sabarubicin did not affect the haemodynamic variables in perfused rat hearts, whereas serious impairments were recorded in the presence of the same concentration of either doxorubicin or epirubicin [54].

3.2
Clinical Development

The first phase I clinical study was carried out in patients with solid tumors to whom sabarubicin was administered, according to the current schedule of antitumor anthracyclines, by a short intravenous infusion given every three weeks. The maximum tolerated dose (MTD) was found to be 100 mg/m^3 [55]. A second study was performed on 24 patients using a weekly schedule. The MTD was determined to be 45 mg/m^2 and recommended intravenous dosages were 30 mg/m^2 and 40 mg/m^2 for three consecutive weeks followed by one week rest in respectively pretreated and naïve patients [56].

The pharmacokinetic behaviour of sabarubicin in human patients was studied using the two treatment schedules outlined above. In one experiment (32 patients) the drug was administered every three weeks, whereas in the other (11 patients) the treatment involved three weekly administrations followed by one week rest. In both experiments the plasma peak concentrations and the area under the concentration vs. time curve showed a linear relationship with the dose. Plasma concentrations were particularly high, in the ranges 0.4–0.6 µg/ml for the lowest (4 mg/m^2) and 9–21 µg/ml for the highest (110 mg/m^2) dose used. No accumulation of the drug was observed in the weekly regimen. The mean elimination half-life was 20.7 h, much shorter than that of doxorubicin or epirubicin and the distribution volume was determined as 95.6 l/m^2 with a standard deviation of 43.4 l/m^2. In fact, a particularly high interpatient variation of pharmacokinetic parameters was found. The small distribution volume, when compared with the other clinically used anthracyclines, is in agreement with the lower rate of uptake in the tissues found in the laboratory animals [47].

On the basis of the relatively better tolerability of sabarubicin, possibly related with a different pharmacokinetic behaviour combined with the higher potency at the target site and of the wider spectrum of antitumor activity when compared with doxorubicin in the preclinical stage, phase II clinical studies have been carried out in patients with different tumor diseases. The compound was shown to be active in non-small cell lung cancer patients with advanced or metastatic disease [57]. Out of 22 evaluable patients, 2 partial responses and 8 minor responses (stable disease) were observed. The drug was well tolerated. In small cell lung cancer, 7 partial responses and 1 stable disease were recorded in a group of 10 patients [58]. Significant response rates were observed in advanced or metastatic platinum/taxane resistant ovarian cancer [59] and in progressive hormone refractory prostate cancer [60].

4
New Sabarubicin Analogues

New analogues of **18** containing different substitutions or configurational modifications have been synthesized and tested in vitro [61]. Compound **24** (4''-*epi*sabarubicin) was obtained upon coupling of 14-acetoxyidarubicinone with *p*-nitrobenzoyl 3-allyloxycarbonyl-4-(*N*,*O*-diallyloxycarbonyl-α,L-acosaminyl)-2-deoxy-α,L-fucoside (**25**), followed by a two step deprotection as described above for the synthesis of **18**. The glycosyl donor **25** was prepared by reaction of phenylthio *N*, *O*-diallyloxycarbonyl-α,L-acosaminide (**26**) with *p*-methoxybenzyl 3-*O*-allyloxycarbonyl-2-deoxy-α,L-fucoside in the presence of IDCP and substitution of the *p*-methoxybenzyl group in the resulting disaccharide derivative with a *p*-nitrobenzoyl residue via cerium ammonium nitrate oxidation and then *p*-nitrobenzoylation.

In order to obtain epimeric **27** and **28**, *p*-methoxybenzyl 4-*O*-acetyl-2,3,6-trideoxy-α,L-*arabino*hexopyranoside (**29**), prepared from di-*O*-acetyl-L-rhamnal by SnCl₄ catalyzed addition of *p*-methoxybenzyl alcohol to give *p*-methoxybenzyl 4-*O*-acetyl-2,3,6-trideoxy-α,L-*arabino*-hex-2-enopyranoside and reduction of the latter with chlorotris(triphenylphosphine)rhodium in anhydrous benzene, was deacetylated and converted, using the Mitsunobu procedure, to the 4-*O*-benzoyl derivative of the L-*lyxo* analogue. The latter was debenzoylated to *p*-methoxybenzyl 2,3,6-trideoxy-α,L-*lyxo*hexopyranoside **30**. Reaction of **30** with **13** or, alternatively, with **26** in the presence of IDCP afforded, respectively, disaccharides **31** and **32** that were converted to the corresponding activated 1-*p*-nitrobenzoyl derivatives and coupled with 14-acetoxyidarubicinone to give, after deprotection as described for the other anthracycline disaccharides, the desired final glycosides.

The concept of elongating the glycosidic portion of the molecule was extended with the preparation of the trisaccharide **34b** starting from **21b** that was converted to the corresponding 1-phenylthio analogue and coupled with **20** in the presence of IDCP to give trisaccharide derivative **33a**. The latter was converted to the activated **33c** via **33b** and used for the glycosylation of 14-acetoxyidarubicinone to give **34a**. Conversion of **34a** to X**34b** was performed in a two step procedure as described for **18** [61]. Compounds **24**, **27** and **28** showed cytotoxic activity in vitro similar to that of **18**, whereas **33b**,

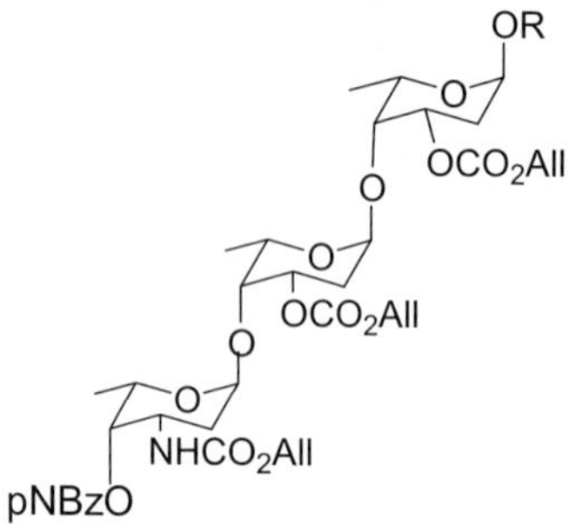

33a (R=pMeOBn)
33b (R=H)
33c (R=pNBz)

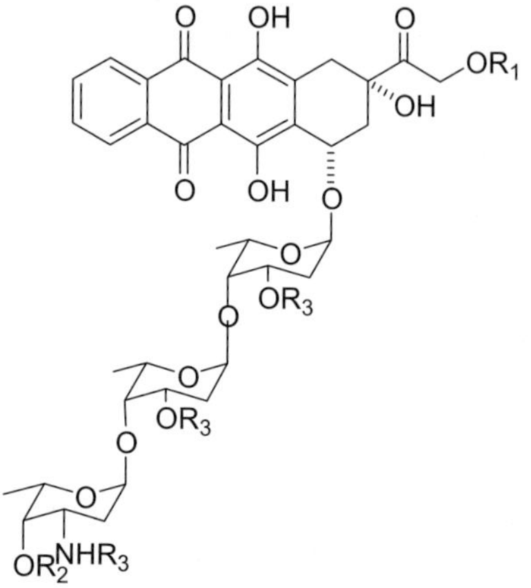

34a (R$_1$=OAc,R$_2$=pNBz, R$_3$=CO$_2$All)
34b (R1=R2=R3=H)

although endowed with topoisomerase II poisoning properties, was much less active, probably because of a lower cell uptake rate.

Analogue **35a** and its 4″-epimer **35b** have also been prepared and tested in comparison with **18**. The compounds were obtained via glycosidation of **22** with **36a** or **36b**, followed by removal of the protecting groups by a deacetylation and a deallylation step. Synthesis of activated disaccharides was performed upon reaction of **20**, in the presence of IDCP, with thiopheyl 3,4-di-allyloxycarboyl-2-deoxy-α,L-fucoside for the preparation of **36a**, or with thiopheyl 3,4-di-allyloxycarboyl-2-deoxy-α,L-rhamnoside for the preparation of **36b**, followed by conversion of the products to the corresponding activated 1-*p*-nitrobenzoyl derivatives as described above for **33c**. The new analogues compared favorably with **18** in animal tumor models, albeit at somewhat higher dosages [62].

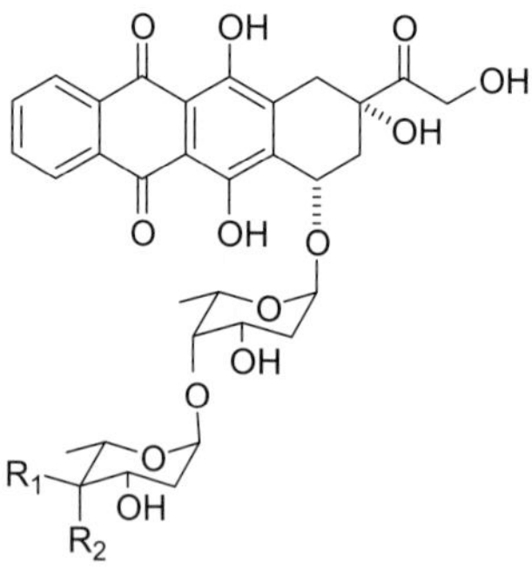

35a (R$_1$=H, R$_2$=OH)
35b (R$_1$=OH, R$_2$=H)

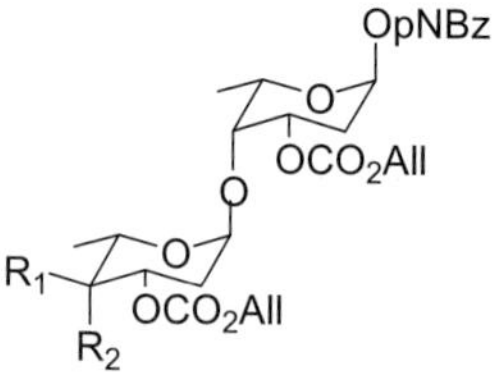

36a (R$_1$=H, R$_2$=OH)
36b (R$_1$=OH, R$_2$=H)

Acknowledgements The discovery and development of sabarubicin was possible through the dedication of the scientists at Menarini Ricerche (Laboratories of Pomezia and Pisa) and at Istituto Nazionale Tumori (Milan) who made this program a reality. The important contributions of G. Ughetto et al. at Istituto di Cristallografia, CNR, Moterotondo, L. Messori et al. at the Department of Chemistry, University of Florence and of the clinical scientists whose names appear in the literature cited, as well as the financial support of Istituto Mobiliare Italiano (IMI) and of Associazione Italiana per la Ricerca sul Cancro (AIRC) are also acknowledged.

References

1. Brockmann H (1963) Forschr Chem Organ Naturst 21:121
2. O'Hagan D (1991) The polyketide metabolites. Ellis Horwood, Chichester UK
3. Arcamone F, Cassinelli G (1998) Curr Med Chem 5:391
4. Fujii I, Ebizuka Y (1997) Chem Rev 97:2511
5. Hutchinson CR (1997) Chem Rev 97:2525
6. Hopwood DA (1997) Chem Rev 97:2465
7. Arcamone F, Di Marco A, Gaetani M, Scotti T (1961) Giorn Microbiol 9:83
8. Arcamone F, Franceschi G, Penco S, Selva A (1969) Tetrahedron Lett 1007
9. Arcamone F (1980) The development of new antitumor anthracyclines. In: Cassady J, Douros J (eds) Anticancer Agents Based on Natural Products Models. Academic, New York, chap 1
10. Arcamone F (1981) Doxorubicin Anticancer Antibiotics. Academic, New York
11. Takeuchi T (1995) J Cancer Res Clin Oncol 121:505
12. Nielsen D, Maare C, Skovsgaard T (1996) Gen Pharmacol 27:255
13. Weiss RB (1992) Semin Oncol 19:670
14. Arcamone F, Animati F, Capranico G, Lombardi P, Pratesi G, Manzini S, Supino R, Zunino F (1997) Pharmacol Ther 76:117
15. Wang AH-J (1992) Curr Opin Struct Biol 2:361
16. Stutter E, Schuetz H, Berg H (1981) Quantitative determination of cooperative anthracycline-DNA binding. In: Cassady J, Douros J (eds) Anticancer Agents Based on Natural Products Models. Academic, New York, p 245
17. Ughetto G (1981) X-Ray diffraction analysis of anthracycline-oligonucleotide complexes. In: Cassady J, Douros J (eds) Anticancer Agents Based on Natural Products Models. Academic, New York, p 296
18. Pullman B (1981) Binding affinities and sequence selectivity in the interaction of antitumor anthracyclines and anthracenediones with double stranded polynucleotides and DNA. In: Cassady J, Douros J (eds) Anticancer Agents Based on Natural Products Models. Academic, New York, p 371
19. Chaires JB, Herrera JE, Waring JM (1990) Biochemistry 29:6145
20. Frederick CA, Williams LD, Ughetto G, van der Marel GA, van Boom JH, Rich A, Wang AH-J (1990) Biochemistry 29:2538
21. Tewey KM, Rowe TC, Yang L, Halligan BD, Liu LF (1984) Science 226:466
22. Capranico G, Zunino F, Kohn KW, Pommier Y (1990) Biochemistry 29:562
23. Capranico G, Kohn KW, Pommier Y (1990) Nucleic Acids Res 22:6611
24. Zunino F, Capranico G (1990) Anticancer Drug Des 4:307
25. Binaschi M, Capranico G, Dal Bo L, Zunino F (1997) Mol Pharmacol 6:1053
26. Gerwitz DA (1992) Biochem Pharmacol 57:727
27. Ishikawa T, Sies H (1984) J Biol Chem 239:3838

28. Formelli F, Zedeck MS, Sternberget SS, Philips FS (1978) Cancer Res 38:3286
29. Ellis CN, Ellis MB, Blakemore WS (1987) Biochem J 245:309
30. Ito H, Miller SC, Billingham ME, Akimoto H, Torti SV, Wade R, Gahlmann R, Lyons G, Kedes L, Torti FM (1990) Proc Natl Acad Sci USA 87:4275
31. Arcamone F, Penco S (1989) Gann Monographs Cancer Res 36:81
32. Capranico G, De Isabella P, Penco S, Tinelli S, Zunino F (1989) Cancer Res 49:2022
33. Capranico G, Supino R, Binaschi M, Capolongo L, Grandi M, Suarato A, Zunino F (1994) Mol Pharmacol 45:908
34. Gresh N, Pullman B, Arcamone F, Menozzi M, Tonani R (1988) Mol Pharmacol 35:251
35. Boivin J, Monneret C, Pais M (1981) Tetrahedron 37:4219
36. El Khadem HS, Matsura D (1982) Carboydr Res 101:C1
37. Horton D, Priebe W, Sznaidman ML, Varela O (1993) J Antibiot 46:1720
38. Borchmann P, Hubel K, Schnell R, Engert A (1997) Int J Clin Pharmacol Ther 35:80
39. Animati F, Berettoni M, Cipollone A, Franciotti M, Lombardi P, Monteagudo P, Arcamone F (1996) J Chem Soc Perkin Trans I:1327
40. Arcamone F, Animati F, Bigioni M, Capranico G, Caserini C, Cipollone A, De Cesare M, Ettorre A, Guano F, Manzini S, Monteagudo E, Pratesi G, Salvatore C, Supino R, Zunino F (1999) Biochem Pharmacol 57:1133
41. Arcamone F, Animati F, Berettoni M, Bigioni M, Capranico G, Casazza AM, Caserini C, Cipollone A, De Cesare M, Franciotti M, Lombardi P, Madami A, Manzini S, Monteagudo E, Polizzi D, Pratesi G, Righetti SC, Salvatore C, Supino R, Zunino F (1997) J Nat Cancer Inst 89:1217
42. Monteagudo E, Madami A, Animati F, Lombardi P, Arcamone F (1997) Carbohydr Res 300:11
43. Ragg F, Mondelli R, Penco S (1988) J Chem Soc Perkin Trans II:1673
44. Cicero DO, Barbato G, Bazzo R (1995) J Am Chem Soc 117:1027
45. Messori L, Temperini C, Piccioli F (2001) Bioorg Med Chem 9:938
46. Temperini C, Messori L, Orioli P, Di Bugno C, Animati F, Ughetto G (2003) Nucleic Acids Res 31:1464
47. Bos AME, De Vries EGE, Dombernovsky P, Aamdal S, Uges DRA, Shuijvers D, Wanders J, Roelwink MWJ, Hanauske AR, Bortini S, Capriati A, Crea AEG, Vermorken JB (2001) Cancer Chemother Pharmacol 48:361
48. Messori L, Piccioli F, Gabrielli S, Orioli P, Angeloni L, Di Bugno C (2002) Bioorg Med Chem 10:3425
49. Pratesi G, De Cesare M, Caserini C, Perego P, Supino R, Bigioni M, Manzini S, Iafrate E, Salvatore C, Casazza AM, Arcamone F, Zunino F (1998) Clin Cancer Res 4:2833
50. Pratesi G, Polizzi D, Perego P, Dal Bo L, Zunino F (2000) Biochem Pharmacol 60:77
51. Bigioni M, Salvatore C, Bullo A, Bellarosa D, Iafrate E, Animati F, Capranico G, Goso C, Maggi CA, Pratesi G, Zunino F, Manzini S (2001) Biochem Pharmacol 62:63
52. Gonzales-Paz O, Polizzi D, De Cesare M, Zunino F, Bigioni M, Maggi CA, Manzini S, Pratesi G (2001) Eur J Cancer 37:431
53. Cirillo R, Sacco G, Venturella S, Brightwell J, Giachetti A, Manzini S (2000) Cardiovasc Pharmacol 35:100
54. Zucchi R, Yu G, Ghelardoni S, Ronca F, Ronca-Testoni S (2000) Brit J Pharmacol 131:342
55. Roelvink M, Aamdal S, Dombernowsky P, Wanders J, Peters S, Bortini S, Crea A, Animati F, Hanauske AR (1999) Eur J Cancer 35(Suppl 4):1161
56. Schrivers D, Bos AME, Dyck J, de Vries EGE, Wanders J, Roelvik M, Fumoleau P, Bortii S, Vermorke JB (2002) Ann Oncol 13:385

57. Tjan-Heijnen VCG, Gerritsen A, ten Valde GPM, Giaccone G, Gamucci T, Comandini A, Capriati A, Mordiva A, Nurmohamed S, Ruland-Adank S (2002) Ann Oncol 13(Suppl 5):148

58. Dickgreber NJ, Welte T, Gillissen A, Dunlop D, Eberhardt W, Wagner T, Swinson D, Capriati A, O'Byrne K (2003) Proc Am Soc Clin Oncol 22:703

59. Caponigro F, Willems P, Sorio R, Floquet A, van Belle S, Demol J, Tambaro R, Comandini A, Capriati A, Adank S, Wanders J (2005) Invest New Drugs 23:85

60. Fiedler W, Tchen N, Bloch J, Fargeot P, Sorio R, Vermorken JB, Collette L, Lacombe D, Twelves C (2006) Eur J Cancer 42:200

61. Cipollone A, Berettoni M, Bigioni M, Binaschi M, Cermele C, Monteagudo E, Olivieri L, Palomba D, Animati F, Goso C, Maggi CA (2002) Bioorg Med Chem 10:1459

62. Bigioni M, Salvatore C, Cipollone A, Berettoni M, Maggi CA, Binaschi M (2005) Lett Drug Design Discovery 2:68

Top Curr Chem (2008) 283: 191–206
DOI 10.1007/128_2007_6
© Springer-Verlag Berlin Heidelberg
Published online: 24 November 2007

Nemorubicin

Massimo Broggini

Istituto di Ricerche Farmacologiche Mario Negri, Via La Masa 19, 20156 Milan, Italy
broggini@marionegri.it

Abstract Nemorubicin is a $3'$-deamino-$3'[2$-(S)-methoxy-4-morpholinyl] derivative of doxorubicin. This derivative has been synthesized in the early 1990s by the Farmitalia Carlo Erba Research Center in Italy. The idea was to develop doxorubicin analogues able to circumvent the emergence of chemoresistance, in particular the multi-drug resistance. The drug was reported to be active in vitro against both murine and human tumor cells resistant to doxorubicin. Similar results were obtained when the drug was administered in vivo to mice bearing multi-drug resistant tumors. The compound retained the same activity also in alkylating agents and topoisomerase II resistant tumors and showed an increased potency relative to the parent drug doxorubicin. It is metabolized via P450 CYP3A enzyme to an extremely cytotoxic derivative. Both nemorubicin and its metabolite have a mechanism of action different from that of doxorubicin, with a key role played by the nucleotide excision repair system. The drug is actively tested in clinics as a single agent or in combination with cisplatin.

Keywords Cytochrome P450 · DNA crosslinks · DNA repair · Resistant cells

Abbreviations

AUC	Area under the curve
NADPH	Nicotinamide adenine dinucleotide phosphate
UV	Ultra violet
DDP	*Cis*-diamminedichloroplatinum(II)

NER   Nucleotide excision repair
BCNU   1,3-Bis(2-chloroethyl)-1-nitroso-urea
NK   Natural killer
CHO   Chinese hamster ovary
mdr   Multi-drug resistance
MRP   Multi-drug resistance related protein
RI   Resistance index
IC50   Concentration inhibiting the growth by 50%
IC70   Concentration inhibiting the growth by 70%
ILS   Increase in lifespan

1
Chemical Properties

Nemorubicin is a doxorubicin derivative characterized by the presence of a 2-S-methoxymorpholinyl group in position 3' of the aminosugar replacing the NH$_2$ group. Its synthesis and chemical characterization was reported in the early 1990s [1, 2]. Nemorubicin is a member of the structural group of morpholinyl anthracyclines whose lead compounds were originally obtained by Takahashi et al. [3] and also by Acton et al. [4] and characterized, especially the cyanomorpholino derivative, as new intensely potent semisynthetic anthracyclines.

Structure 1

The methoxymorpholinyl group is responsible for the high lipophilicity which facilitates its entrance into the cells. In fact, in several human and murine-derived cancer cell lines, equal concentrations of doxorubicin or nemorubicin resulted in a more than four times higher intracellular levels of nemorubicin than doxorubicin [1]. As a freeze dried powder, the drug presents a remarkable stability remaining unaltered at temperatures as high as 45 °C for at least three months [5].

The in vitro cytotoxicity experiments performed in different murine and human derived cancer cell lines, showed that nemorubicin was more potent than doxorubicin being able to induce a 50% inhibition of growth at concentrations three times lower than those necessary with doxorubicin [1].

Nemorubicin was extremely active in vivo when administered intravenously, intraperitoneally, or by oral route [6]. It produced antitumor activity similar to that of doxorubicin but at approximately 100 times lower concentrations. The differences in potency observed between in vitro and in vivo experiments strongly indicated that the drug is transformed to an extremely potent metabolite.

2
Metabolism and Pharmacokinetics

2.1
Pharmacokinetics in Animals

Using ^{14}C-labeled nemorubicin, Breda et al. investigated the pharmacokinetics, the excretion and the urinary metabolites of the drug in rats and dogs [7]. Similar pharmacokinetic profiles were obtained in male and female rats indicating a sex-independent disposition. In the first 8 hours following administration, 75–100% of total radioactivity measured was associated with unchanged nemorubicin. If the area under the curve (AUC) values are considered, nemorubicin represented more than 50% of the total radioactivity. The 13-OH metabolite was present in relatively low concentrations, accounting for approximately 10% of the total levels. The plasma profile of nemorubicin showed a rapid distribution phase followed by a slower phase of elimination.

In dogs the distribution was similar, again with a rapid distribution phase and a slower elimination phase, but the levels of unchanged nemorubicin were already at early time points (stages) only 1/5 of the total radioactivity. In this case the levels of the 13-OH metabolite were higher than those of nemorubicin in terms of AUCs.

In the first four days following administration, approximately 70% of the dose was recovered in urine and feces, with the latter accumulating the majority of the drug. In a way the results were similar in rats and dogs. Using radio TLC, the authors determined that 13-OH metabolite accounted for 20–25% of the drug excreted in the urine with other metabolites present in lower concentrations.

2.2
In Vitro and In Vivo Metabolism Studies

Preincubation of nemorubicin with human liver microsomes in the presence of nicotinamide adenine dinucleotide phosphate (NADPH) increased by 50-fold the in vitro cytotoxicity of the drug against human ovarian carcinoma-derived cells ES-2 [8]. This increase in potency was abolished by co-incubation with cyclosporin A, a substrate of cytochrome P450 3A. Moreover, a difference in the kind of DNA damage was found, with nemorubicin being able to induce single and double strand breaks in the absence of microsomes and shifting to the formation of DNA crosslinks when incubated in the presence of activated microsomes.

The involvement of CYP3A in the metabolism of nemorubicin was further demonstrated by experiments in which microsomes from mice treated with the CYP3A inducer pregnenolone-16alpha-carbonitrile were used [9]. Incubation with induced microsomes resulted in a 2.5-fold increase in cytotoxic activity of nemorubicin compared to that obtained incubating the drug with microsomes obtained from untreated mice. Conversely, microsomes obtained from mice treated with the CYP3A inhibitor troleandomycin showed a lower ability to activate nemorubicin in terms of cytotoxicity. In addition, this effect was also observed in vivo. Using the murine ovarian tumor model M5076, Quintieri et al. found that pretreatment of mice with troleandomycin completely abolished the antitumor and antimetastatic activity of nemorubicin [9].

Detailed analysis of the products generated by incubating nemorubicin with activated microsomes obtained from different species has been published [10]. Human, monkey, dog and rat microsomes are all able to transform nemorubicin, although with different metabolic rates. From human microsomes, the authors isolated eight different fluorescent metabolites not present when the incubation was carried out in the absence of NADPH. Three of them were more polar than nemorubicin and one was identified as nemorubicinol, the 13-OH derivative of nemorubicin, thanks to the availability of authentic standards. Similarly of the five less polar metabolites, two corresponded to available standards relative to PNU-156686 and PNU-159682 (the 3'-deamino-3'',4'-anhydro-[2''(S)-methoxy-3''(R)-oxy-4''-morpholinyl]doxorubicin).

Experiments performed using cDNA-expressed CYP3A enzymes identified the human CYP3A4, the major isoform of CYP3A enzyme in human liver, as the most active enzyme responsible for the activation of nemorubicin. CYP3A1 and CYP3A2 rat enzymes were also able to induce the activation of the drug. Conversely, human CYP3A5 and CYP3A7 were not able to significantly transform nemorubicin [11].

PNU-159682 was also identified as a P450-dependent metabolite of nemorubicin by reversed-phase HPLC coupled to ultra violet (UV) and radiometric detection and by LC-MS/MS in other independent experimental conditions [12]. Kinetic analysis of the formation of this metabolite showed that the reaction was linear over a range of concentration of the substrate indicating that a single microsomal enzyme or different enzymes with a similar apparent Km are able to catalyze the conversion of nemorubicin to PNU-159682. To confirm previous indication of the enzyme(s) responsible for this biotransformation, the authors elegantly demonstrated that CYP3A4 was indeed the

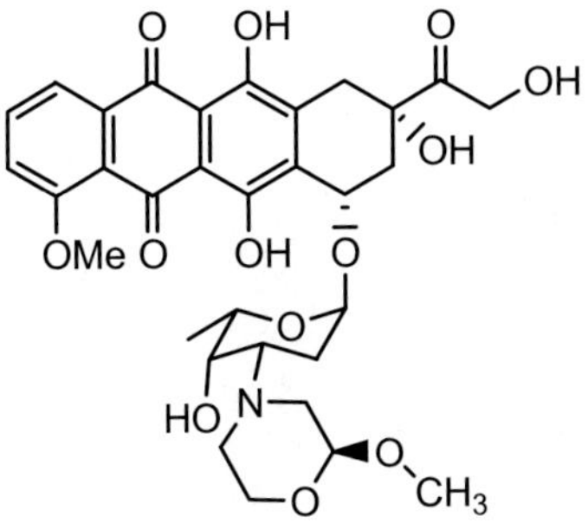

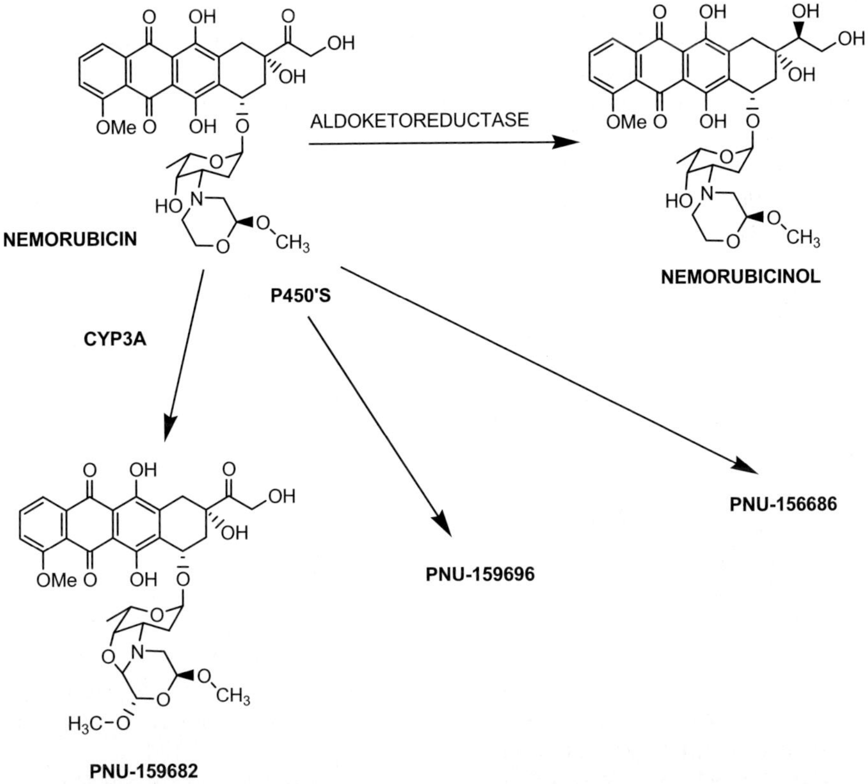

Scheme 1 Metabolic conversion of nemorubicin through the action of aldoketoreductases and P450 enzymes. The structure of the major metabolites, PNU-159682 and nemorubicinol are reported

enzyme responsible for this transformation. Microsomal preparations containing single recombinant human cytochrome P enzymes were in fact used to follow the conversion of nemorubicin to PNU-159682 (see Fig. 1) [12]. CYP3A4 expressing preparations effectively transformed nemorubicin, while

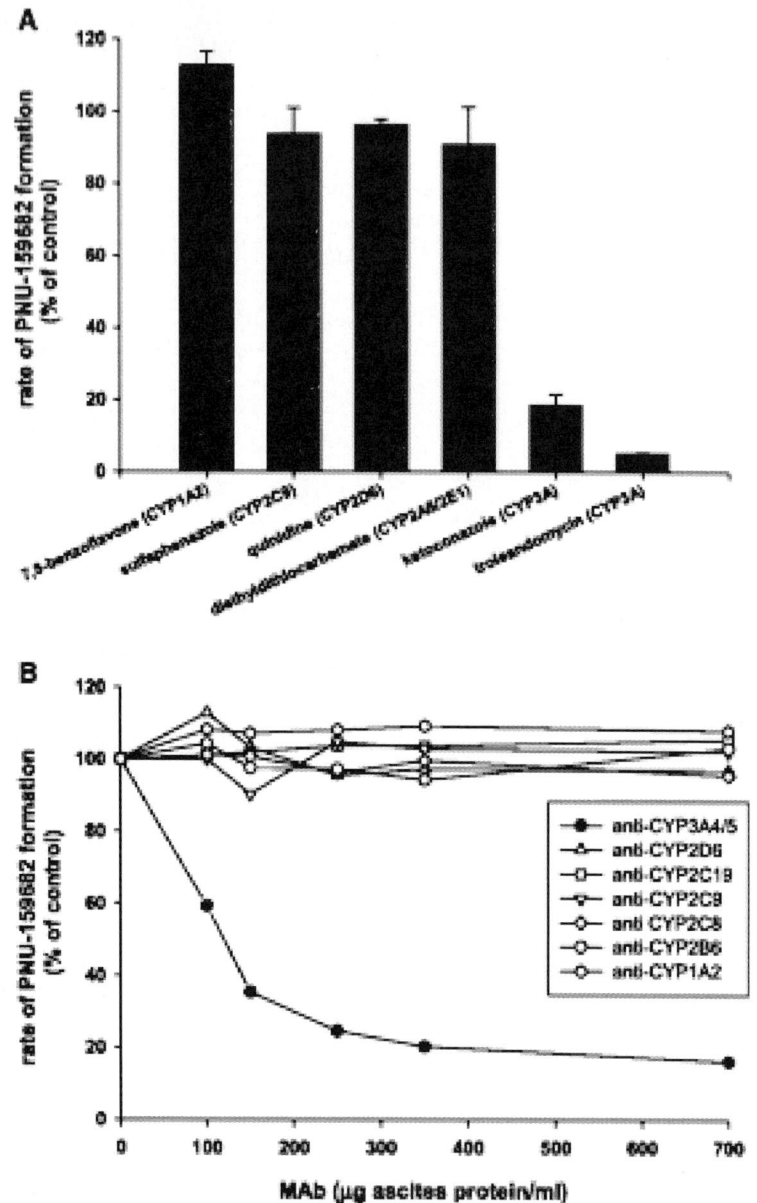

microsomes containing CYP1A2, CYP2A6, CYP2B6, CYP2C8, CYP2C9*1, CYP2C19, CYP2D6*1, CYP2E1, CYP3A5, were not able to produce detectable amounts of PNU-159682.

The metabolic conversion of nemorubicin, with the structures of the major metabolites are reported in Scheme 1.

◄ **Fig. 1** Effect of specific CYP inhibitors (*panel A*) or specific anti-CYP monoclonal antibodies (*panel B*) on the conversion of nemorubicin to its metabolite PNU-159682. *Panel A*: all incubations were carried out in the absence (control) or presence of appropriate concentrations of chemical inhibitors, at 37 °C for 10 minutes, using pooled human liver microsomes (0.25 mg ml^{-1}), nemorubicin (20 µmol l^{-1}), and NADPH (0.5 mmol). Values represent percentage of control activity, and are the mean ± standard error of three independent experiments. *Panel B*: pooled human liver microsomes (20 pmol of total CYP), preincubated in the absence or presence of increasing amounts of monoclonal antibodies at 37 °C for 5 minutes, were incubated with nemorubicin (20 µmol l^{-1}) and NADPH (0.5 mmol) at 37 °C for 10 minutes. Values represent percentage of control activity determined in the absence of monoclonal antibodies, and are the mean of duplicate determinations. Adapted from [12]

2.3
Pharmacokinetics in Humans

In a phase I study conducted on patients receiving nemorubicin as bolus injection at dose levels ranging from 675 to 2250 µg/m^2, a two exponential phase equation described the plasma decay of unchanged nemorubicin [13]. In this study a HPLC method with fluorescence detection was employed. The plasma levels of the 13-OH metabolite were always near the limit of quantization of the assay and hence no reliable pharmacokinetic parameters could be determined. The elimination half life of nemorubicin was on the order of 40 hours. A relatively high clearance (approximately 600 ml min^{-1}/m^2) was found with no dependence on the dose administered. This suggests that linear kinetics is operating, at least in the range of doses utilized. Similarly, the AUC values were proportional to the dose administered again suggesting linearity in the range of doses tested. Finally, the values of the volume of distribution were indicative of a strong distribution into tissues.

In a phase II study with iv bolus injection of nemorubicin, plasma elimination half life was estimated in 49 hours, a value similar to that reported in the above-mentioned phase I study [14]. The authors also evaluated the AUC values of the 13-OH metabolite which accounted for approximately 10% of the unchanged nemorubicin.

Pharmacokinetic analysis of nemorubicin in plasma of patients receiving the drug orally in a phase I study was evaluated [15]. Patients received the drug at doses ranging from 59 to 940 µg/m^2. Plasma levels could be measured starting from the dose of 470 µg/m^2. The C_{max} for the parent compound was reached after approximately 4 hours. A 13-OH metabolite was detectable in 6 out of 9 patients and its levels were half of those of nemorubicin.

Interestingly, a relation between the dose administered and the C_{max} and experimental AUC values was found. The elimination half life was very variable among the patients with a mean value of 11 hours and a range of 3.1 and 36.8 hours.

When the drug was given as intravenous infusion of 3 hours, plasma levels could be measured in 12 patients at dose levels of 1000, 1250 and 1500 µg/m^2 [16]. C_{max} values were around 2 ng ml^{-1} with AUC values increasing with the dose administered. As for the bolus injection, rapid clearance and high volume of distribution were observed. The urinary excretion determined in the first 72 hours was low with values accounting for approximately 2.5% of the total administered dose. The 13-OH metabolite was also scarcely excreted in the urine with values reaching 2.3% of the total administered dose.

Intrahepatic artery injection, at doses ranging from 200 to 800 µg/m^2 in patients with hepatocellular carcinoma gave long terminal half lives (61–98 hours) and high volume of distribution (1400–2300 l/m^2) [17].

3
Cytotoxic Activity

3.1
In Vitro Activity

Nemorubicin was active in vitro against several murine and human cancer cell lines, always showing higher potency than doxorubicin. Comparison of the intracellular concentrations reached after exposure to the same amount of nemorubicin and doxorubicin in different cancer cell lines showed that nemorubicin uptake was faster but was also always present in higher amounts.

In a detailed analysis of the in vitro activity of nemorubicin against human leukemia and lymphoma cell lines, Kuhl et al. found that in all the 14 human cell lines examined, nemorubicin was more active (5 to 7 times) than doxorubicin [18]. In a panel of human hepatocellular carcinoma-derived cell lines, nemorubicin was more than ten times more potent than both doxorubicin and epidoxorubicin [19].

The activity of nemorubicin against a panel of human cancer cell lines was compared with the activity against hematopoietic progenitors [20]. Cells derived from human bone marrow or from human umbilical cord blood were used as the source of normal progenitor cells. The cells isolated from these different sources were equivalent in terms of sensitivity to both nemorubicin and doxorubicin. The comparison of the activity of nemorubicin and doxorubicin in cancer and normal cells was therefore analyzed in detail using cells derived from umbilical cord blood, due to the reduced availability of human bone marrow. Both drugs showed a superior activity against cancer cells than against normal progenitors. The ratio between the concentration inhibiting the growth by 70% (IC70) obtained in the two systems (the higher the better) was 4.5 for doxorubicin and 5.3 for nemorubicin.

A more favorable activity against normal cells was found for nemorubicin when its activity was tested in isolated perfused heart and compared with that of doxorubicin [21]. Using equimolar concentrations of doxorubicin or nemorubicin the authors found that doxorubicin induced a prolongation of the $S\alpha T$ segment and of the $Q\text{-}F_{max}$ interval together with a reduction in dF/dt_{max} and coronary flow. On the contrary, nemorubicin did not cause these effects, the only alteration shared with doxorubicin being the reduction in dF/dt_{max}.

The cardiotoxicity exerted by doxorubicin and nemorubicin was compared in vivo in rats receiving acute or chronic treatments [22]. Acute treatment with doxorubicin was associated with electrocardiogram changes, including enlargement of QRS complex and $S\alpha T$ segment, and alteration of hemodynamic parameters while corresponding doses of nemorubicin (retaining the same antitumor activity of the dose of doxorubicin used in these experiments) gave only an increase in the QRS complex duration.

Chronic treatment of doxorubicin gave anorexia and diarrhea associated with reduction in body weight compared to untreated rats or to rats receiving equiactive doses of nemorubicin. In terms of cardiotoxicity, doxorubicin induced severe cardiomyopathy with damage detectable at the histological level consisting of vacuolation and myofibrils loss. At the electrocardiogram level, doxorubicin induced a progressive widening of the $S\alpha T$ segment and an increase in T wave. Nemorubicin did not show any significant alteration indicating a markedly less cardiotoxicity in vivo compared to doxorubicin.

3.2
Activity Against Resistant Cells

Being structurally related to doxorubicin, nemorubicin was initially tested for its activity against resistant cells presenting the multi-drug resistance (mdr) phenotype. Interestingly, nemorubicin retained its activity against a panel of

Table 1 Concentration inhibiting the growth by 50% (IC50, $ng\,ml^{-1}$) of doxorubicin and nemorubicin in cells sensitive or resistant to doxorubicin due to overexpression of mdr-1 or MRP

Cell lines	Doxorubicin	RI	Nemorubicin	RI
LOVO/Doxorubicin	2180	36	33	2
LOVO	60		16	
GLC4/Doxorubicin	3010	91	0.020	2.5
GLC	33		0.008	
MCF7/Doxorubicin	1230	123	17	1.3
MCF-7	10		13	
HL60/Daunorubicin	718	194	27.2	8.0
HL60	8.6		3.4	

RI resistance index is the ratio between IC50 on resistant and sensitive cells

cancer cell lines having a high degree of resistance to doxorubicin due to the expression of mdr or multi-drug resistance related protein (MRP) [1, 2, 18, 23–25]. Table 1 summarizes the results obtained in a pair of cells sensitive and resistant to doxorubicin.

Nemorubicin showed also activity against cell lines selected for resistance to agents acting with a mechanism of action different from that of doxorubicin, including platinum derivatives, alkylating agents and topoisomerase I and II inhibitors [18, 23, 26].

Overall these data suggest a broader spectrum of activity of nemorubicin both in sensitive and resistant tumors.

3.3
In Vivo Activity

The difference in potency between nemorubicin and doxorubicin was also demonstrated in all the in vivo experimental systems tested. Nemorubicin, in fact, produced antitumor activity similar to that of doxorubicin at doses approximately 100 times lower. This is likely to be due to the formation of a more potent metabolite via the P450 CYP3A isoenzyme, as already discussed.

Confirming the evidences presented in vitro, nemorubicin retained the antitumor activity in vivo against cells resistant to different drugs including anthracyclines, platinum derivatives, and alkylating agents. Nemorubicin showed appreciable activity even when administered orally. Marked activity was found against human hepatocellular carcinomas xenografted in nude mice. Nemorubicin activity against hepatic metastasis was studied using the model of murine reticular cell sarcoma M5076 [27]. This model in fact preferentially metastatizes to the liver. Differently from doxorubicin, nemorubicin, injected intravenously, was more active in reducing liver metastasis than in reducing primary tumor. This relative preference for liver metastasis was confirmed by oral administration of the drug.

The antitumor activity of the synthetic metabolite PNU-159682 was tested in vivo using two different model systems: the murine L1210 leukemia and the human mammary carcinoma xenograft MX-1. In the L1210 model, PNU-159682 given intravenously as a single dose of $15\,\mu g\,kg^{-1}$ gave an increase in life span (ILS) of 29%, similar to that produced by a single injection of nemorubicin (ILS 31%) at the dose of $90\,\mu g\,kg^{-1}$.

PNU-159682 showed, in the MX-1 model sensitive to nemorubicin but refractory to doxorubicin, high activity at doses as low as $4\,\mu g\,kg^{-1}$ with 4 out of 7 mice presenting complete response defined as absence of palpable tumor.

To increase nemorubicin delivery, new ways of administrations have been explored. Nemorubicin encapsulated in IL2-activated natural killer (NK) cells has been used in a model of hepatic metastasis. Injection of nemorubicin encapsulated in activated NK cells was much more active in inhibiting the formation of liver metastasis than the same amount of drug given as free drug. Encapsulated nemorubicin gave the same activity obtained by using approximately twice as many doses of free drug. Interestingly, with this formulation 9 out of 10 mice did not show evidence of metastasis when sacrificed 21 days following tumor injection. The same dose given as free drug resulted in only 1 out of 10 mice without metastasis. Moreover, this formulation did not result in myelosuppression which was evident when the equiactive amount of nemorubicin was administered as free drug.

Combination of doxorubicin with alkylating agents was tested in murine leukemia models [28]. Nemorubicin was administered iv followed 24 hours after by treatment with cis-diamminedichloroplatinum(II) (DDP) or mitomycin C given intravenously. In this tumor model (L1210) nemorubicin alone, at the highest non-toxic dose gave a 67% ILS. DDP and mitomycin C showed values of ILS of 50% and 0%, respectively. Combination of DDP and nemorubicin gave a 133% ILS while nemorubicin plus mitomycin C was able to induce a 93% ILS. These results indicate that these combinations are synergistic and indeed provided the rational, together with the in vitro mechanistic studies, of the use of combination between DDP and nemorubicin in the clinic.

4
Mechanism of Action

To get further insights in the nemorubicin mechanism of action, a large amount of data has been published on experiments performed on a murine L1210 leukemia subline made resistant to nemorubicin. These results confirm that nemorubicin acts through a mechanism of action different from that of classical anthracyclines. In fact, the nemorubicin resistant cell line showed similar responsiveness, compared to the parental cell line it is derived from, to agents such as doxorubicin, vinblastine, or campthotecin, and shows even an increased sensitivity to agents such as DDP, melphalan, 1,3-bis (2-chloroethyl)-1-nitroso-urea (BCNU), mitomycin C, and 5-fluorouracil. Interestingly enough, this pattern of sensitivity was also retained in in vivo experiments (Table 2).

In addition, when tested for its sensitivity to UV light, the nemorubicin resistant cell line was found to be more sensitive to UV than the parental line. Analysis of the ability of parental and nemorubicin resistant cells to repair a UV damaged plasmid (host cell reactivation assay) identified a defect in DNA repair ability in the nemorubicin resistant cells. Since UV damage is mostly repaired by nucleotide excision repair (NER), this particular DNA repair system is likely to be relevant for the mechanism of action of nemorubicin. Indeed experiments performed in isogenic cell lines derived from Chinese hamster ovary (CHO) cells confirmed the previous findings [29].

Table 2 Comparison of the in vivo activity of nemorubicin and doxorubicin in different tumor models

Tumor model	Nemorubicin		Doxorubicin		Refs.
	Dose ($\mu g\,Kg^{-1}$)	Tumor growth inhibition (%)	Dose ($\mu g\,kg^{-1}$)	Tumor growth inhibition (%)	
M5076	50	90	6000	93	[34]
BEL-742	50	58.8	5500	56.4	[16]
Zip-177	50	45.4	5500	47.7	[16]
MX-1	70	99	6600	72	[3]
MTV	65	94	5850	99	[1]
3LL	100	36	7500	85	[1]
N592	85	89	6000	55	[3]
A549	45	29	4000	31	[3]
CX-1	45	9	5200	34	[3]
LoVo	45	43	4000	83	[3]

*M*5076 Murine reticular cell sarcoma, *BEL* – 742 Human hepatocellular carcinoma, *Zip* – 177 Human hepatocellular carcinoma, *MX* – 1 Human mammary carcinoma, *MTV* Murine mammary carcinoma, *3LL* Murine lung carcinoma, *N*592 human small cell lung carcinoma, *A*549 human lung adenocarcinoma, *CX* – 1 human colocarcinoma, *LoVo* human colocarcinoma

The antiproliferative effect of nemorubicin was evaluated by colony assay on CHO cells wild type (AA8) and in two sublines defective in defined steps of the NER pathway (UV 96 for ERCC1 and UV 61 for ERCC6). NER proficient cells were more sensitive to nemorubicin than the two NER defective sublines, with a three to four-fold difference in the concentration needed to inhibit the growth by 50%. Restoration of NER activity in UV 96 subline (through transfection of the human ERCC1 gene), rendered the cells sensitive as the parental ones to the drug.

This finding was further corroborated in another pair of cells derived from the murine L1210 leukemia cells; the subline defective in NER was 2 to 3 times more resistant to treatment with nemorubicin than the parent cell line [30].

It must be noted that this behavior is quite peculiar, shared so far only by another anticancer agent of marine origin, ET-743, while the majority of the DNA damaging anticancer agents are in general more sensitive in cells with defects in NER than in those with a normal DNA repair ability [31].

In the A2780 model of human ovarian cancer cells growing in vitro, treatment for 24 hours with nemorubicin induced a block of cells in the G2 phase of the cell cycle with an increase in the number of cells present in the subG1 phase, which is an indication of apoptosis induction [32]. The potent metabolite PNU-159682 showed instead a cell cycle block more evident in the S phase, again associated with the presence of a subG1 peak. When the treatment was performed for shorter times (1 hour), nemorubicin still induced

Table 3 In vitro and in vivo activity of different anticancer agents against parental and nemorubicin resistant L1210

Compound	In vitro RI	In vivo activity (% ILS)	
		L1210	L1210/Nemorubicin
Nemorubicin	8.5	63	0
Doxorubicin	0.9	100	108
Vinblastine	1.2	56	63
Camptothecin	1.3	94	125
Melphalan	0.7	111	> 567
BCNU	0.7	294	267
5-Fluorouracil	0.5	75	72
Mitomycin-C	0.3	56	> 650
DDP	0.5	75	> 567

RI resistance index is the ratio between IC50 on resistant and sensitive cells

an arrest of cells in G2 phase, which was however reverted after 72 hours. PNU-159682, on the contrary, in the same treatment conditions, induced an irreversible block and all the treated cancer cells were apoptotic after 72 hours [32]. The detailed comparison with doxorubicin, performed using bromodeoxyuridine pulse and chase experiments [32], showed that while doxorubicin affected cells in S phase at the moment of treatment, nemorubicin and PNU-159682 were effective, independent of the cell cycle phase in which the cells were present.

5
Clinical Studies

In a phase I study using intravenous bolus injection with 21 day intervals [13] nemorubicin was given at doses ranging from 30 to 2250 $\mu g/m^2$.

Reversible myelotoxicity, with a delayed nadir in comparison with doxorubicin was the dose limiting toxicity. At these dose levels, no evidence of cardiotoxicity was observed. Other side effects were nausea and vomiting and transient increase in hepatic transaminases. Even if this was a phase I study and clinical response was not the primary end point, four objective responses were recorded. Of these, one was a complete response, one a partial response and two were minor responses. This phase I study identified the maximum tolerated dose in 1500 $\mu g/m^2$, with a recommended dose for phase II studies of 1250 $\mu g/m^2$.

Different treatment schedules were tested in other phase I studies. A bolus intravenous administration given daily for three days every 28 days was tested in a way to reduce nausea and vomiting [33]. While this schedule of treatment showed indeed lower non-hematological toxicities compared to the single-dose study, reversible myelosuppression was again the dose limiting toxicity. The recommended dose for phase II was fixed at 500 µg/m^2/day ×3. Of the thirty patients treated with this schedule, no one showed partial or complete responses.

Prolonged intravenous infusion of 3 hours every 28 days was tested in 14 patients [16] with a dose range from 1000 to 15 000 µg/m^2. No clinical responses were observed. An increase in myelotoxicity, compared to the bolus intravenous injection was found, with comparable non-hematological toxicities. These results suggest that prolonged infusion does not seem to be a better way to administer nemorubicin to patients.

An interesting possibility offered by the physicochemical properties of the molecule is oral administration of the drug. The safety of this treatment modality has been tested in a phase I study in which 21 patients were enrolled [15]. The schedule consisted of a treatment performed every 4 weeks at doses ranging from 59 to 940 µg/m^2. Nemorubicin was given as hard gelatine capsules. As for the intravenous route, the main hematological toxicity was neutropenia which was found to be highly variable among patients. Only 1 out of 21 patients experienced a grade 4 toxicity and the maximum tolerated dose for myelotoxicity was 940 µg/m^2. Non-hematological toxicity was mostly nausea and vomiting, which was very severe and defined the maximal tolerated dose in 820 µg/m^2. No responses were found with this treatment modality. Due to the high interpatient variability and to the severe and prolonged gastrointestinal side effects, the oral treatment was not further developed.

A phase I has been conducted in patients with unresectable hepatocellular carcinoma administering nemorubicin via the hepatic artery [17]. Nemorubicin was administered at doses of 200, 400, 600 and 800 µg/m^2 every 8 weeks. Thirteen patients received 25 cycles of treatment with moderate and reversible toxicity. At the doses studied, no grade 4 (which is the most severe) toxicities were observed and the dose limiting toxicity was not reached. An increase in transaminases was observed starting from 400 µg/m^2. Based on these findings, the maximum tolerated dose was established at 800 µg/m^2, where reversible transaminites of grade 3 were observed in 2 out of 3 patients. Six partial responses were achieved.

Based on these encouraging results, a phase II/III trial was started in China using nemorubicin at 600 µg/m^2 every 6 weeks administered via intrahepatic artery in patients with unresectable hepatocellular carcinoma previously untreated [34]. In the 24 patients where efficacy was evaluable, partial responses were obtained in 5 (20.8%). Moderate toxicity was observed in these treatment conditions with moderate reversible increase in transaminases. Further analysis of nemorubicin efficacy in hepatocellular carcinoma showed that in 57 patients treated 11 showed complete or partial response and 17 had a stabilization of the disease lasting more than three months [35].

Based on the result obtained at the preclinical level, a phase II trial of the combination of nemorubicin and DDP has begun in patients with hepatocellular carcinoma [35]. The trial is still ongoing and the results not yet available.

6
Concluding Remarks

Nemorubicin, although structurally related to doxorubicin, has a mechanism of action different from that of classical anthracyclines. Several features make nemorubicin an interesting drug deserving further clinical development:

- It retains antitumor activity in vitro and in vivo in cells resistant to different anticancer agents including anthracyclines, topoisomerases inhibitors and alkylating agents.
- It is able to overcome multi-drug resistance associated with decreased intracellular uptake.
- It has no cardiotoxic effects and is generally well tolerated.
- Cells selected for resistance to nemorubicin show collateral sensitivity to alkylating agents and platinum derivatives.
- When combined with DDP it shows synergistic antitumor activity without increase in toxicity.

The clinical responses observed in phase I/II trials encourage its further clinical development either as single agent or in combination with currently used anticancer agents.

References

1. Grandi M, Pezzoni G, Ballinari D, Capolongo L, Suarato A, Bargiotti A, Faiardi D, Spreafico F (1990) Cancer Treat Rev 17:133
2. Grandi M, Ballinari D, Capolongo L, Pastori A, Ripamonti M, Suarato A, Spreafico F (1991) Haematologica 76(Suppl 3):181
3. Takahashi Y, Kinoshita M, Masuda T, Tatsuta K, Takeuchi T, Umezawa H (1982) J Antibiot 35:117
4. Acton EM, Tong GL, Mosher CW, Wolgemuth RL (1984) J Med Chem 27:638
5. Farina A, Quaglia MG, Doldo A, Calandra S, Gallo FR (1993) J Pharm Biomed Anal 11:1215
6. Ripamonti M, Pezzoni G, Pesenti E, Pastori A, Farao M, Bargiotti A, Suarato A, Spreafico F, Grandi M (1992) Br J Cancer 65:703
7. Breda M, Benedetti MS, Battaglia R, Castelli MG, Poggesi I, Spinelli R, Hackett AM, Dostert P (2000) Pharmacol Res 41:239
8. Lau DH, Duran GE, Lewis AD, Sikic BI (1994) Br J Cancer 70:79
9. Quintieri L, Rosato A, Napoli E, Sola F, Geroni C, Floreani M, Zanovello P (2000) Cancer Res 60:3232

10. Beulz-Riche D, Robert J, Menard C, Ratanasavanh D (2001) Fundam Clin Pharmacol 15:373
11. Lu H, Waxman DJ (2005) Mol Pharmacol 67:212
12. Quintieri L, Geroni C, Fantin M, Battaglia R, Rosato A, Speed W, Zanovello P, Floreani M (2005) Clin Cancer Res 11:1608–1617
13. Vasey PA, Bissett D, Strolin-Benedetti M, Poggesi I, Breda M, Adams L, Wilson P, Pacciarini MA, Kaye SB, Cassidy J (1995) Cancer Res 55:2090
14. Bakker M, Droz JP, Hanauske AR, Verweij J, van Oosterom AT, Groen HJ, Pacciarini MA, Domenigoni L, van Weissenbruch F, Pianezzola E, de Vries EG (1998) Br J Cancer 77:139
15. Sessa C, Zucchetti M, Ghielmini M, Bauer J, D'Incalci M, de Jong J, Naegele H, Rossi S, Pacciarini MA, Domenigoni L, Cavalli F (1999) Cancer Chemother Pharmacol 44:403
16. Fokkema E, Verweij J, van Oosterom AT, Uges DR, Spinelli R, Valota O, de Vries EG, Groen HJ (2000) Br J Cancer 82:767
17. Sun Y, Li Z, Sun L, Valota O, Battaglia R, Pacciarini MA (2003) Proc Am Soc Clin Oncol 22:1448
18. Kuhl JS, Duran GE, Chao NJ, Sikic BI (1993) Cancer Chemother Pharmacol 33:10
19. Yuan S, Zhang X, Lu L, Xu C, Yang W, Ding J (2004) Anticancer Drugs 15:641
20. Ghielmini M, Colli E, Bosshard G, Pennella G, Geroni C, Torri V, D'Incalci M, Cavalli F, Sessa C (1998) Cancer Chemother Pharmacol 42:235
21. Danesi R, Agen C, Grandi M, Nardini V, Bevilacqua G, Del Tacca M (1993) Eur J Cancer 29A:1560
22. Pinna A, Agen C, Di Paolo A, Innocenti F, Nardini D, Danesi R, Del Tacca M (1994) J Environ Pathol Toxicol Oncol 13:25
23. van der Graaf WT, Mulder NH, Meijer C, de Vries EG (1995) Cancer Chemother Pharmacol 35:345
24. Bakker M, Renes J, Groenhuijzen A, Visser P, Timmer-Bosscha H, Muller M, Groen HJ, Smit EF, de Vries EG (1997) Int J Cancer 73:362
25. Michieli M, Damiani D, Michelutti A, Melli C, Ermacora A, Geromin A, Fanin R, Russo D, Baccarani M (1996) Haematologica 81:295
26. Alvino E, Gilberti S, Cantagallo D, Massoud R, Gatteschi A, Tentori L, Bonmassar E, D'Atri S (1997) Cancer Chemother Pharmacol 40:180
27. Geroni C, Quintieri L, Valota O, Rosato A, Zanovello P, Floreani M, Sola F, Pacciarini MA (2002) Eur J Cancer 38:S19
28. Geroni C, Pacciarini MA, Valota O (2002) Eur J Cancer 38:S20
29. Geroni C, Sabatino MA, Ballinari D, Ciomei M, Marsiglio A, Quintieri L, Floreani M, Broggini M (2006) Proc Am Assoc Cancer Res 47 [Abstract 3845]
30. Sabatino MA, Geroni C, Broggini M (2006) Eur J Cancer Suppl 4:152
31. Damia G, Silvestri S, Carrassa L, Filiberti L, Faircloth GT, Liberi G, Foiani M, D'Incalci M (2001) Int J Cancer 92:583
32. Albanese C, Geroni C, Ciomei M (2006) Proc Am Assoc Cancer Res 47 [Abstract 4666]
33. Ratain MJ, Skoog LA, O'Brien SM, Cooper N, Schilsky RL, Vogelzang NJ, Gerber M, Narang PK, Nicol SJ (1997) Ann Oncol 8:807
34. Sun Y, Yang J, Luo P, Zhang Y, Yan Y, Sun L, Pacciarini MA, Valota O, Geroni C (2004) Eur J Cancer Suppl 2:143
35. Pacciarini MA, Geroni C, Sabatino MA, Ciomei M, Valota O, Ballinari D, Capolongo L, Broggini M (2006) J Clin Oncol 24:14116

Author Index Volumes 251–283

The volume numbers are printed in italics

Bargon J, see Kuhn LT (2007) *276*: 125–154

Barigelletti F, see Flamigni L (2007) *281*: 143–204

Barthel BL, see Koch TH (2008) *283*: 141–170

Bayly SR, see Beer PD (2005) *255*: 125–162

Beck-Sickinger AG, see Haack M (2007) *278*: 243–288

Beer PD, Bayly SR (2005) Anion Sensing by Metal-Based Receptors. *255*: 125–162

Beretta GL, Zunino F (2008) Molecular Mechanisms of Anthracycline Activity. *283*: 1–19

Bergamini G, see Balzani V (2007) *280*: 1–36

Bergamini G, see Campagna S (2007) *280*: 117–214

Bertini L, Bruschi M, de Gioia L, Fantucci P, Greco C, Zampella G (2007) Quantum Chemical Investigations of Reaction Paths of Metalloenzymes and Biomimetic Models – The Hydrogenase Example. *268*: 1–46

Bier FF, see Heise C (2005) *261*: 1–25

Blum LJ, see Marquette CA (2005) *261*: 113–129

Boiteau L, see Pascal R (2005) *259*: 69–122

Bolhuis PG, see Dellago C (2007) *268*: 291–317

Borovkov VV, Inoue Y (2006) Supramolecular Chirogenesis in Host–Guest Systems Containing Porphyrinoids. *265*: 89–146

Boschi A, Duatti A, Uccelli L (2005) Development of Technetium-99m and Rhenium-188 Radiopharmaceuticals Containing a Terminal Metal–Nitrido Multiple Bond for Diagnosis and Therapy. *252*: 85–115

Braga D, D'Addario D, Giaffreda SL, Maini L, Polito M, Grepioni F (2005) Intra-Solid and Inter-Solid Reactions of Molecular Crystals: a Green Route to Crystal Engineering. *254*: 71–94

Bräse S, see Jung N (2007) *278*: 1–88

Braverman S, Cherkinsky M (2007) [2,3]Sigmatropic Rearrangements of Propargylic and Allenic Systems. *275*: 67–101

Brebion F, see Crich D (2006) *263*: 1–38

Breinbauer R, see Mentel M (2007) *278*: 209–241

Breit B (2007) Recent Advances in Alkene Hydroformylation. *279*: 139–172

Brizard A, Oda R, Huc I (2005) Chirality Effects in Self-assembled Fibrillar Networks. *256*: 167–218

Broene RD (2007) Reductive Coupling of Unactivated Alkenes and Alkynes. *279*: 209–248

Broggini M (2008) Nemorubicin. *283*: 191–206

Bromfield K, see Ljungdahl N (2007) *278*: 89–134

Bruce IJ, see del Campo A (2005) *260*: 77–111

Bruschi M, see Bertini L (2007) *268*: 1–46

Bur SK (2007) 1,3-Sulfur Shifts: Mechanism and Synthetic Utility. *274*: 125–171

Burkhart DJ, see Koch TH (2008) *283*: 141–170

Campagna S, Puntoriero F, Nastasi F, Bergamini G, Balzani V (2007) Photochemistry and Photophysics of Coordination Compounds: Ruthenium. *280*: 117–214

Campagna S, see Balzani V (2007) *280*: 1–36

del Campo A, Bruce IJ (2005) Substrate Patterning and Activation Strategies for DNA Chip Fabrication. *260*: 77–111

Capobianco ML, Catapano CV (2008) Daunomycin-TFO Conjugates for Downregulation of Gene Expression. *283*: 45–71

Cardinali F, see Armaroli N (2007) *280*: 69–115

Carney CK, Harry SR, Sewell SL, Wright DW (2007) Detoxification Biominerals. *270*: 155–185

Subject Index

Printing: Krips bv, Meppel, The Netherlands
Binding: Stürtz, Würzburg, Germany